# A Practical Approach to Industrial Systems Integration

Industry 4.0 and Industrial Internet of Things: Cases of Manufacturing, Energy, Building, Environment and Business Data Integration Using Ethernet and OPC Technologies

## Tom Wanyama
McMaster University, Hamilton, Ontario

A Practical Approach to Industrial Systems Integration

International Automation Academy
ISBN: 978-0-9948503-0-0

# TABLE OF CONTENTS

# *Preface*

## Industrial System Integration and Industry 4.0

Industry 4.0 is associated with the seamless integration of hardware, software and technology-based services, to enhance manufacturing productivity and improve efficiency. The backbone of this integration is industrial networks and technologies that support data integrations. Modern industrial networks enable companies to collect information automatically from manufacturing management systems, government agencies, utility companies, buildings and environment management systems. This information is filtered, and integrated using advanced data analytics systems to produce knowledge that is used to control and manage manufacturing production. Ultimately, industrial networks enable Industry 4.0 to have the following benefits: improved safety, increased uptime, lower energy costs, and improved maintenance; all of which lead to manufacturing competitiveness in cyber-physical production systems supported by Smart Grid implementations.

Most legacy industrial networks were designed for data communication among sensors, actuators, remote I/O modules, and controllers. This data tends to be smalls, in many cases less no more than 32 bytes per message. At the plant level of the industrial network architecture, devices share large amounts of information, which necessitates high capacity networks. Therefore, most plant level networks use Ethernet based protocols such as PROFINET and Ethernet/IP, since Ethernet is known to carry large amount of data. Moreover, using Ethernet based protocols at the plant level makes it easy to integrate plant and business networks, a feature required by Industry 4.0. It should be noted that Ethernet is becoming ubiquitous and cost effective, with common physical links and increased speed and determinism.

The author appreciates the role played by Ethernet based protocols in integrating production, logistics and customer service centers of companies. He also believes that the use of Ethernet based networks in integrating plant devices from plant level to the sensor and actual level shall continue to grow. Therefore this book covers the following network protocols: Standard TCP/IP Ethernet, Ethernet/IP, BACnet, and IEC61850, PROFINET, and Modbus TCP.

## Scope and Features

This book is a result of lecture notes that the author has developed and refined over many years to cover the core concepts of industrial networks in depth, and to provide general overview of advanced concepts that students may pursue in advanced courses such as those on communication protocols and communication devices development. The book has the following features that set it aside from any competing textbook on industrial networks:

*Industry 4.0 manufacturing Paradigm:* This book introduces Industry 4.0 manufacturing paradigm, and how it affects industrial automation systems. It also describes the role played by Ethernet in supporting industry 4.0. In doing so, the book shows how various seemingly unrelated technologies are covered in a single book. In the final chapter, a recommended approach for designing Industry 4.0 solutions is presented, together with three low cost industry 4.0 implementation examples.

*Book Style:* The author wrote this book in a style that supports the principle of integrating lectures, laboratories and course project, also known as integrated learning. According to this paradigm, the book presents the theory of the networks through lecture sized chapters, on the other hand network configuration and communication programming are covered through laboratories. The course project is used to integrate the work covered in lectures and laboratories through an experiential learning paradigm. Some of the laboratories in this book are available on line on request from the author.

*Discussion Questions:* The author provides discussion questions at the end of every chapter. Some of these questions are open ended, designed to spur discussion that go beyond the material presented in this book.

*Diagrams:* The author teaches in an engineering technology program and most of his students are visual. He has therefore developed many visual representations of major concepts of Industrial networks. These diagrams help to illustrate the various concepts described in the book.

*Laboratories:* Eight laboratories in the appendices are a key feature of this book. They act as a bridge between theory and practice. If you have access to the equipment used in our laboratories, then you can use the labs to gain hands on experience. If you have access to similar but different equipment, then you can use our labs as a guide to developing your own labs that will help you to bridge theory and practice. But if you do not have access to any equipment, our labs will help you to understand how theory and practice complement each other. The laboratories on OPC technology are available for remote access on request.

## Summary

The material covered in this book fall under the general field of Internet of Things. The book focuses on accessing and integrating manufacturing, energy, building, environment and business data using Ethernet and OPC technologies. To support experiential learning, the book presents six laboratories and one project. The project can be done by accessing the author's lab equipment remotely through the Internet.

## Acknowledgements

This text, A Practical Approach to Industrial Systems Integration, is an outgrowth of my lecture notes I have used in the course PROCTECH6AS3: Advance System Components and Integration of the Process Automation Technology program in the School of Engineering Practice and Technology (SEPT) at McMaster University, Hamilton, Ontario, Canada. The text has benefited greatly from the feedback I have received from many of my curious and keenly observant students. I sincerely give thanks to all of them. I also thank Dr. Ishwar Singh of the School of Engineering Practice and Technology (SEPT) at McMaster University for his support during the development of the laboratories presented in this book. Last but not least, I would like to thank my wife Grace Kentaro Wanyama for her support and for the hours she spent designing formatting this book.

# Chapter 1 *Introduction*

Automation systems can be integrated at many different levels, such as machine level, plant level, company level, or industry level. At machine level the focus of integration is the selection of machine components, interfacing the components, and programming the controllers, while at the company or industry level the focus is sharing information so as to optimize associated processes.

## 1.1    Legacy Automation Systems

Generally, an automation system is a communication system consisting of controllers and links. The structure of the system usually reflects the structure of the plant. Ideally, each unit of the plant has its own controller, interacting with the controllers of the other, related units, and mirroring their physical interaction. While the applications of automation systems differ widely, there is little difference in their overall architecture. The small differences stem from domain requirements that need to be embedded in the systems. Such requirements include:

- Need for *-proof devices (* {water, weather, explosion})
- Availability (24hours operation and hot repair)
- Regulations (e.g. Food and Drug Administration, energy efficiency)
- Standards (IEC and ISA)
- Tradition and customer relationships

Figure 1.1 shows the conceptual model (structure and relationships) of building, manufacturing/production and electrical substation automation system architectures. The figure reveals that traditionally, integration of automation systems focuses on the primary processes (e.g. vertical integration of manufacturing systems), leaving out other value chain components that indirectly affect production and product quality, such as energy, materials, logistics and customer service. Furthermore, Figure 1.1 shows that the vertical integration of automation systems involves five levels, namely: sensor, field, control/plant, supervisory, and execution and enterprise.

1. Sensor level (Level 1): This level deals with data acquisition (Sensors) and commands that modify the process (Actuators). The level interacts with the primary technologies (Level 0) of the automation system.

2. Field level (Level 2): This level handles data transmission. There is no processing at this level, except measurement correction and built-in protection. Devices at this level include: remote IOs, fieldbus/sensor, and fieldbus/actuator interfaces.

3. Control level (Level 3): This level has two components, namely: Unit (cell) control and group (area) control. Unit control provides the following functions to part of a group control:

   - Measure: Sampling, scaling, processing, calibration.

   - Control: regulation, set-points and parameters

   - Command: sequencing, protection and interlocking

4. The group (area) control is part of the control level that offers the following functions to a well-defined part of the plant:

   - Coordinate individual subgroups

   - Adjust set-points and parameters

   - Command several units as a whole

5. Supervision level (Level 4): This level supervises production, carries out site optimization, and executes operations. In addition, the level provides the following functions to the automation system: plant visualization, data log and storage, and historical data access. This is the SCADA level of a plant.

6. Enterprise and execution level (Level 5):  For small-single site organizations, this level is usually implemented as a single unit. But for large or multisite organization, the enterprise component of this level may be used to handle corporate issues, while the execution component is used to deal with the running of each site and reporting to the enterprise. Generally, the enterprise deals with administration, finances, human resources, documentation, and long-term planning. It also set production goals, plans the enterprise and resources, coordinate different sites, and manage orders. On the other hand, the execution covers the management of the production execution plan, resources, workflow, quality supervision, production scheduling, and maintenance.

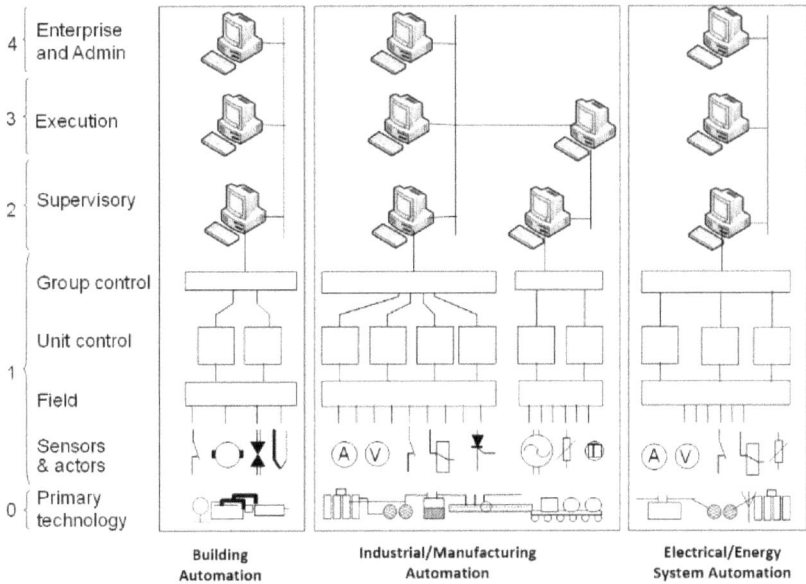

**Figure 1.1: Conceptual Model of the Automation System Architecture**

The ANS/ISA standard 95 categorizes the levels of automation systems shown in Figure 1.1 into three groups, namely: Enterprise Resource Planning (level 4 in Figure 1.1) which deals with business activities such as business planning, logistics, production scheduling and operational management; Manufacturing Execution System (level 3 in Figure 1.1) which covers manufacturing activities such as manufacturing operations and control, production dispatching, detailed production scheduling, and reliability assurance; and finally Control and Command System that deals with process control.

Figure 1.2 shows the description of the most important elements of automation system architecture. It shows that if a plant has multiple primary processes, is intrgated only at the enterprise or execution level using the business logic. The similarities among automation systems could be leveraged to increase efficiencies and optimize skills usage so as to reduce production and training costs. Moreover information from one subsystem could be used in real-time to modify or improve the performance of another. For example electricity pricing and Kilo-Volt- Ampere (KVA) demand information could be used to modify production scheduling so as to flatten the demand curve and reduce the KVA maximum demand of plants,

[3]

which ultimately reduces production cost; instead of waiting for the energy bill and using the business logic later to modify the behavior of the production process.

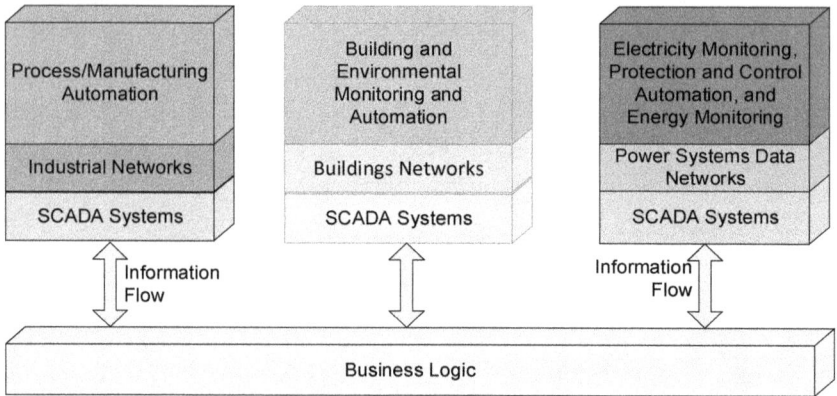

**Figure 1.2: Important Components of the Automation System Architecture**

Figure 1.3 shows the mapping of functionalities of legacy automation systems architecture onto the system hardware and software components. The figure reveals that legacy automation systems have the following general characteristics:

- They view materials and energy as inputs to processes as opposed to parameters that can be measured and used to optimize production.
- Their sensor data is used by the controller to control a predefined or unknown process model.
- Sophisticated systems log some controller, process, and sensor data and send it to Enterprise Resource Planning (ERP) and Manufacturing Execution Systems (MES), where it is analyzed to produce information used for decision making.
- The outputs of process models are used to command actuators that modify the transformation process to produce process products.
- They do not make any effort to optimize waste and environmental impact.

[4]

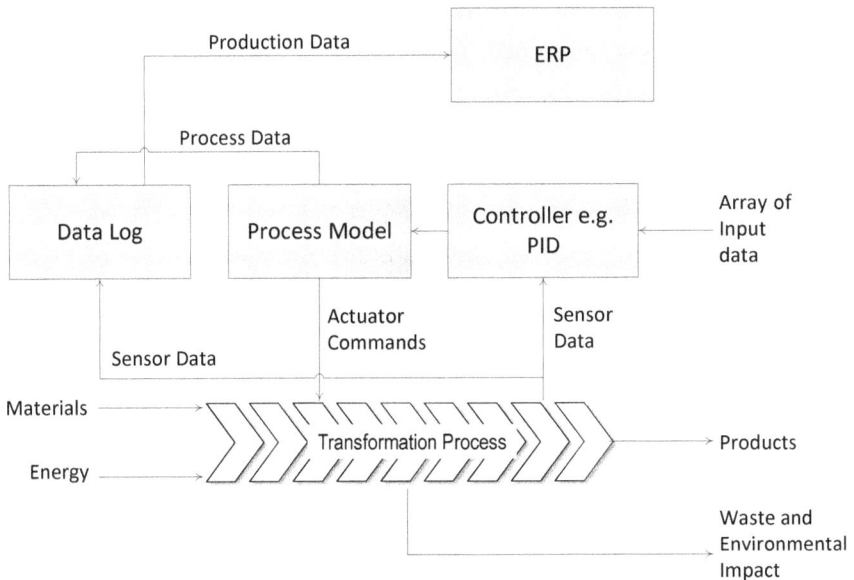

**Figure 1.3: Legacy Approach to Process Automation**

## 1.2 Modern Automation Systems

Due to developments in Ethernet network technology, it is possible to integrate all aspects of production, including those that fall under different production processes, such as manufacturing, logistics, energy, and business logic. This ensures plant and companywide optimization, and makes it easy to transfer knowledge and skills from one area of automation system type to others. Figure 1.4 shows the conceptual model of the modern automation system architecture.

The figure is based on the fact that various Ethernet based communication protocols that were designed for specific industries, such as Ethernet IP for manufacturing and process automation, BACnet for building automation, Standard Ethernet TCP/IP for business networks, and IEC61850 for electricity substation automation share the same physical layer (Ethernet) of the OSI reference model. This makes it possible to integrate their data into a single ERP or SCADA system. Once data is contained in a single system, it becomes possible to use information from one industry to optimize processes in another industry. An example of such a scenario is the use of energy and environmental information to optimized production in the manufacturing industry.

Figure 1.5 shows the description of the most important elements of modern automation systems architecture. Generally, the figure reveals that automation data can be transferred to cloud servers through the business network using various web technologies. This data can then be analyzed and integrated with data from other value chain sources, such as regulatory organizations and material suppliers, to generate corporate-wide intelligence. The ability to integrate various value chain organizations has resulted into a new manufacturing paradigm generally referred to in literature as Industry 4.0 or Digital Industry. This paradigm seeks to leverage the potential optimization in production and logistics caused by increased and integrated industrial automation, intelligent system monitoring, and autonomous decision-making that is supported by real-time or almost real-time communication at all levels.

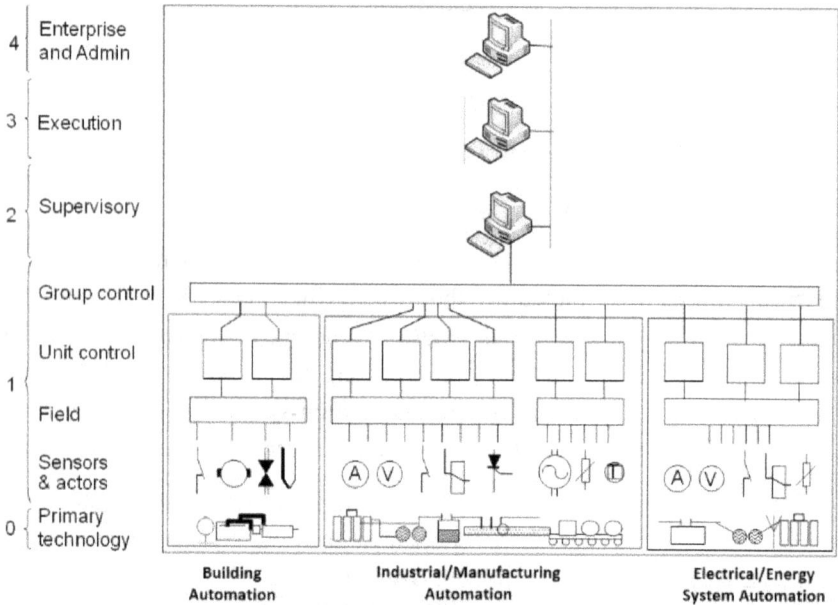

**Figure 1.4: Conceptual Model of the Modern Automation System Architecture**

Figure 1.6 shows Industry 4.0 mapping of automation systems architecture onto hardware and software components of modern automation systems. The figure shows that Industry 4.0 automation systems utilize information about the quality and type of materials; information from energy utilities (smart grid), waste and environment management systems, as well as information from other value chain

[6]

participants to optimize production processes through advanced data analytics and intelligent process and control models. Again, this is only possible because of the ability to integrate data from different flavors of Ethernet based networks used in various industries that affect manufacturing.

Figures 1.5 and 1.6 show that Industry 4.0 approach to process/manufacturing automation generates and utilizes a lot of distributed data, making it necessary to implement cloud computing solutions. This involves deploying groups of remote servers and software networks allowing online access to computer services and resources such as data storage, advanced data analytics, process and systems monitoring, and data access and visualization. Increased access to computing power availed to industry through cloud computing enables extensive simulations of different aspects of industrial operation, and allows processing of huge amounts of data that industry already collects but does not use adequately. Taken together, this should result in better and faster decision making. Figure 1.6 shows that Industry 4.0 uses Artificial Intelligence methods to improve decision making as well as process and manufacturing control models. Moreover, the Figure shows that besides cloud computing resources, organizations need to address the following factors when implementing Industry 4.0:

- Global Database: Industry 4.0 relies on information sharing among production and logistics centers of companies. To ensure that all the centers utilize the most current information, it is necessary to receive information from a single location, defined by Schuh et al [2] as the "single source of truth" and presented in this paper as the global company database. Since Industry 4.0 companies generate and utilize vast amounts of data, the global database is beyond what can be integrated into user devices. Therefore, it is usually implemented as part of cloud computing services.

- Automation: With the help of appropriate information processing units, automation based Industry 4.0 enables a bank of machines to perform tasks independently. Automation systems link the virtual information/digital environment and the physical world by integrating information processing units with sensors/actuators. To realize full Industry 4.0 benefits, companies should maximize the automation of processes.

- Networks: In order for different production centers to share information, network nodes must be physically interconnected through data communication networks. This allows massive amounts of information (generated by industrial sensors) to be shared and sorted using various types of communications protocols. Industry 4.0 standard utilizes TCP/IP, the most common internet protocol, to link several billion devices.

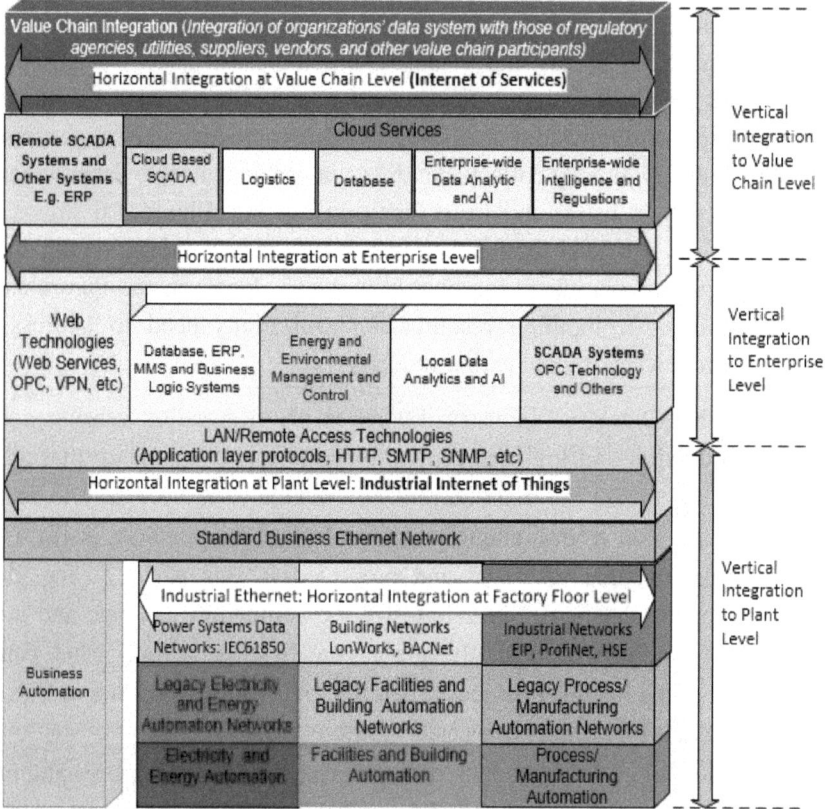

**Figure 1.5: IIoT Approach to Process/Manufacturing Automation**

It should be noted that two additional enablers of Industry 4.0 have been largely overlooked by the literature, namely, education/training and reliable smart grid:

- *Education/Training:* Advances in manufacturing have to be supported by access to a highly skilled, educated workforce of technicians, trades workers and engineers. As the fourth industrial revolution takes hold, it is important that educational institutions modify their programs and develop new industrial network courses producing Industry 4.0 ready graduates.
- *Smart Grid:* The power industry is both a beneficiary and an enabler of Industry 4.0. From a benefits perspective, smart grid deployment permits shifting its production model from consumption-oriented generation to generation-optimized consumption, with the latter model increasing efficiency and lowering the need to invest in expensive generation capacity. On the other hand, the power industry is expected to support Industry 4.0 by supplying cost effective and reliable energy.

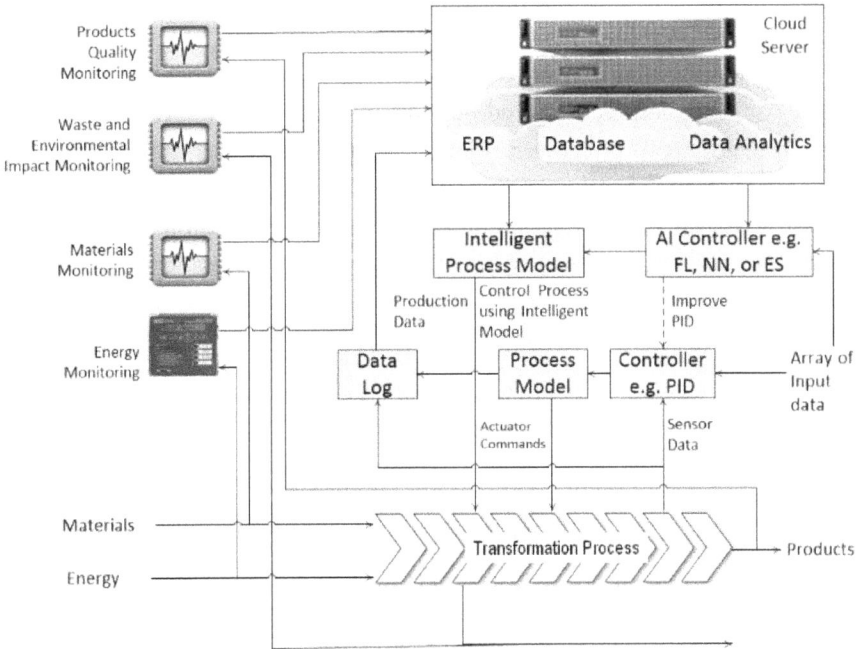

**Figure 1.6: Industry 4.0 Approach to Process Automation**

This book focuses on supporting the networks and data integration enabler of Industry 4.0. Moreover, the book indirectly supports the Education/Training enabler.

## 1.3    System Components and Integration

In order to support the paradigm of Industry 4.0, integrated manufacturing has to be considered as made of up of the following components:

- Manufacturing process control and management.
- Environmental and building management.
- Energy management.
- Logistics, quality control, and customer service management.
- Value chain organizations.

Integrating these components result in what is generally referred to in literature as Industrial Internet of Things (IIoT). While the aspects associated with the above manufacturing components are usually addressed, normally they are taken to be part of different automation processes (see Figures 1.1 and 1.2). Yet, information from all these aspects has to be collected and integrated automatically in real-time or almost real-time in order to achieve increased automation, as well as the benefits of Industry 4.0. Therefore, network technologies such as Ethernet that enable the electronic integration of all aspects of manufacturing are the backbone of Industry 4.0. They support both vertical and horizontal integration of automation systems as shown in Figures 1.4 and 1.5. Figure 1.1 and Figure 1.2 show that vertical integration allows the sharing of data among all levels of automation systems. On the other hand, horizontal integration enables sharing of data among difference primary processes in the plant. Moreover, horizontal integration of automation systems supports sharing of information among value chain participant of Industry 4.0 through cloud servers and the Internet (See Figures 1.4, 1.5 and 1.7).

This book is based on the modern automation model shown in Figure 1.6. It covers the network technologies that support system integration of process/manufacturing automation, building automation and environment management, as well as energy management and electricity systems automation (smart grid systems). In other words, this book covers the network infrastructure of IIoT. The book clearly specifies components of network technologies that are common to all industries, and the components that are specific to applications in particular industries. Moreover, the book covers the development of Supervisory Control And Data Acquisition (SCADA) systems based on Open Productivity and Control (OPC) technology. It shows that OPC based SCADA systems are not only applicable in different automation industries, but can also be used to integrate data from those industries.

[10]

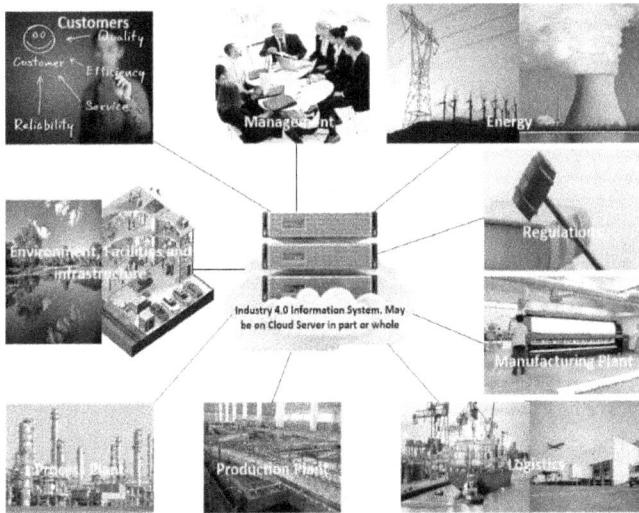

**Figure 1.7: Sharing of Information among Value Chain Organizations**

This book prepares students to work in Industry 4.0 environment and to support companies to realize the benefits of Industry 4.0. The book is designed to support experiential learning using the following laboratory work (see Appendices):

Figure 1.8 illustrates the main components of the laboratory set up used to carry out one of the Ethernet IP labs in this book. The main objective of the laboratory is to teach students the process of identifying useable IP addresses as well as assigning addresses to the network nodes. After assigning IP addresses to the network nodes and ensuring that all nodes communicate correctly, students move on the second part of the laboratory that involves configuring the communications in the PLC programming software. In this laboratory, Automation Direct Productivity 3000 software is used. Figure B.10 presents the configuration for communication between the Productivity 3000 PLC and the Eaton ELC-CAENET remote I/O module. Since the communication between the remote I/O and the PLC is based on the principle of implicit messaging of Ethernet IP, memory is allocated automatically to the input, output, and configuration data. Therefore, there is no need to program the communication, instead data is read and written automatically. Explicit messaging is used for communication between Automation Direct Productivity 3000 PLC and Eaton PowerXL DG1 VFD. Therefore, in this

[11]

lab students have to configure a client just as they do for implicit messaging, but the reading and writing of data is implemented through explicit messages. Figure B.13 shows a message instruction for reading data from the Eaton VFD (see Appendices A and B).

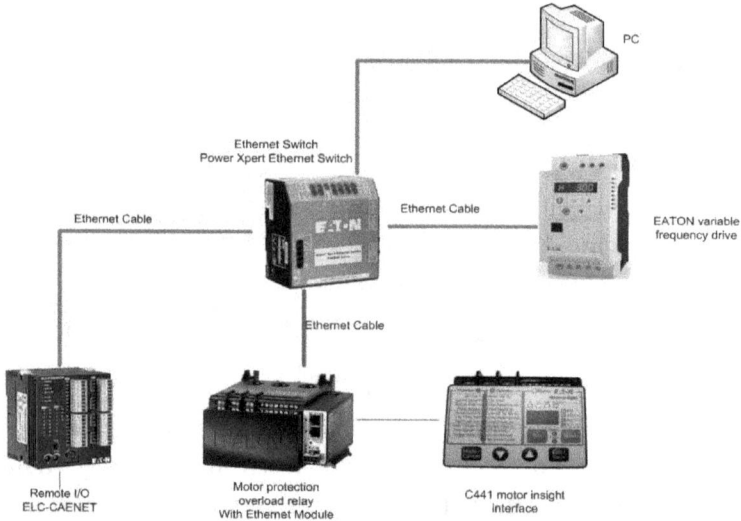

**Figure 1.8: Ethernet IP laboratory setup**

- The IEC61850 electrical substation automation is taught using laboratories based on the Schweitzer Engineering Laboratories SEL751A relay. The first laboratory focuses on the configuration of the metering and communication parameters of the relay as well as programming the relay I/Os. The second laboratory covers Generic Object Oriented Substation Event (GOOSE) messaging and the third lab deals with Manufacturing Messaging Specification (MMS) (see Appendices C and D).

- We have developed a laboratory for teaching building automation programming, as well as networking using BACnet, including OPC data access. The laboratory based on PXC controller and it can be expanded to cover various flavours of BACnet, including BACnet/IP, BACnet/Ethernet, and BACnet MS/TP. Moreover, it can be modified to address the integration of BACnet and Modbus protocols (see Appendices G and H).

[12]

- SCADA is taught using OPC DataHub and KEPserver. The focus of OPC is to integrate data from manufacturing automation, electrical substation automation and monitoring, and building automation. SCADA laboratories are based on a piece of equipment that has a Micrologix 1400 PLC for manufacturing automation, SEL751A relay for electricity/energy control and monitoring, and the PXC controller for building automation.

Figure 1.9 shows the equipment for carrying labs associated with the network infrastructure of IIoT. The equipment is accessible over the university VPN, which enables remote access to the PLC, BACnet building automation controller, as well as an electricity feeder protection relay data. The equipment is also used to support the data integration project. The project is an extension of the OPC laboratories. It requires students to carry out the following tasks:

- Complete the HMI that is started in one of the labs (Appendix F).
- Integrate manufacturing, electricity/Energy, and building automation data.
- Move data from OPC client to Excel, MatLab, or SQL database using Dynamic Data Exchange (DDE).

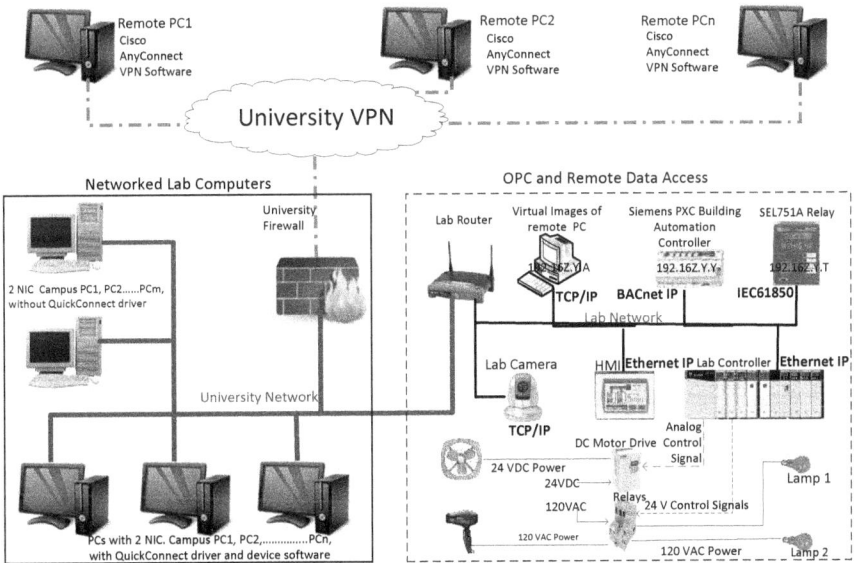

**Figure 1.9: Network Infrastructure of IIoT Lab Equipment**

[13]

## 1.4    References

[1] Benno Bunse, Henning Kagermann, and Wolfgand Wahlster, Industry 4.0: Smart Manufacturing for the Future, Germany Trade and Invest, Berlin, German, July 2014. Available as of April 12, 2015 from http://www.gtai.de/GTAI/Content/EN/Invest/_SharedDocs/Downloads/GTAI/Brochures/Industries/industrie4.0-smart-manufacturing-for-the-future-en.pdf

[2] Günther Schuh, Christina Reuter, Annika Hauptvogel ,and Christian Dölle, Hypotheses for a Theory of Production in the Context of Industrie 4.0, Advances in Production Technology, Lecture Notes in Production Engineering , C. Brecher (ed.), 11 pp., 2015, ISBN 978-3-319-12303-5

[3] Günther Schuh, Till Potente, Cathrin Wesch-Potente, Anja Ruth Weber , Jan-Philipp Prote, "Collaboration Mechanisms to increase Productivity in the Context of Industry 4.0", in Proc. 2nd CIRP Robust Manufacturing Conference, Katja Windt (ed.), (Bremen, Germany; 7 – 9 July 2014) , 51 pp., 2014,ISBN: 978-1-63439-362-1.

[4] Ishwar Singh, Nafia Al-Mutawaly, and Tom Wanyama, "Teaching Network Technologies that Support Industry 4.0", in Proc. CEEA Canadian Engineering Education Conf., CEEA2015, (Hamilton, Ontario; 31 Many- 03 June 2015), 2015.

## 1.5    Discussion Questions

**Question 1**

How can data integration be used to improve manufacturing efficiency? Hint: use the electricity and manufacturing data as examples.

**Question 2**

Explain security challenges faced by systems integration and how they can be alleviated.

**Question 3**

Describe Artificial Intelligence methods that can be used in data analytics and in process control.

**Question 4**

Explain the importance of vertical and horizontal integration in Industry 4.0 enabled manufacturing facilities.

# Chapter 2 *Legacy Industrial Networks*

Historically, industrial networks or network protocols have been developed by different companies and later became standards. Manufacturers of industrial automation equipment continue to implement many of these protocols in devices for the following reasons:

- Protocols are supported by big automation companies.
- Industrial devices that use legacy network protocols tend to have long useful life.
- Legacy industrial network protocols exhibit a high degree of determinism with low cycle times, typically less than 15 milliseconds.

## 2.1 Automation Systems Hierarchy and Communication Protocols

Most legacy industrial networks were designed for data communication among sensors, actuators, remote I/O modules, and controllers. This data tends to be captured as short strings (small data) with many cases no more than 32 bytes per message. On the other hands, there is need for vertical integration of automation systems (See figures 1.1 and 1.2) to enable data and information sharing at all levels of automation systems. Figure 2.1 present automation system levels as well as automation systems components classification based on the ANS/ISA standard 95. The standard defines terminology and good practices for manufacturing information technologies and systems integration. The functions of these levels are described in Section1.2. Moreover, the figure shows that as one move up the levels of automation systems, communication requirements of the system incrementally change from high speed and determinism to large data handling and information integrity, resulting in using different communication protocol at different levels of the system, as well as a hieratical industrial network architecture shown in Figure 2.3.

Figure 2.2 shows some network protocols that are appropriate for data communication at various levels of automation systems. From the plant/control level of the industrial network architecture onwards, devices share massive information, which requires high capacity networks. Therefore, most networks at those levels use Ethernet based protocols such as PROFINET and Ethernet/IP (Figure 2.2 and Figure 2.3), because Ethernet is capable of carrying large amounts of data.

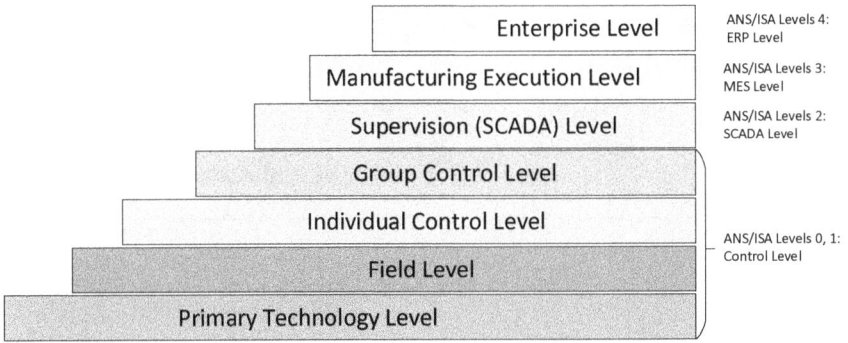

**Figure 2.1: Plant Automation Levels and ANS/ISA Standard 95 Hierarchy**

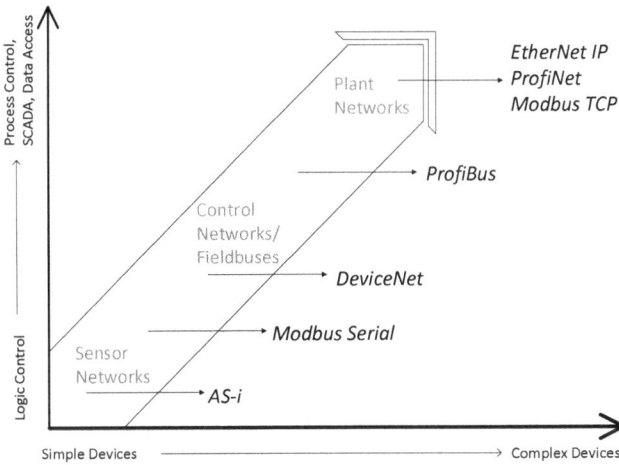

**Figure 2.2: Appropriate Network Protocol for Automation Levels**

In addition, Ethernet based protocols at the plant level make it easy to integrate plant networks and business networks, a feature desired by Industry 4.0. As Ethernet is ubiquitous and cost effective, with common physical links, high speed and determinism, it is poise to becomes the de facto protocol for industrial networks. It is expected that the data flow between process level sensors/actuators and plant data centers will continuously grow. In line with this growth, this book focuses on adopting Ethernet based protocols as a means of integrating manufacturing, energy, environmental and buildings monitoring, supply chain and customer service centers. Therefore, the book covers Ethernet protocol topics

[17]

including: Standard TCP/IP Ethernet, Ethernet IP, BACnet, IEC61850, PROFINET, Modbus TCP.

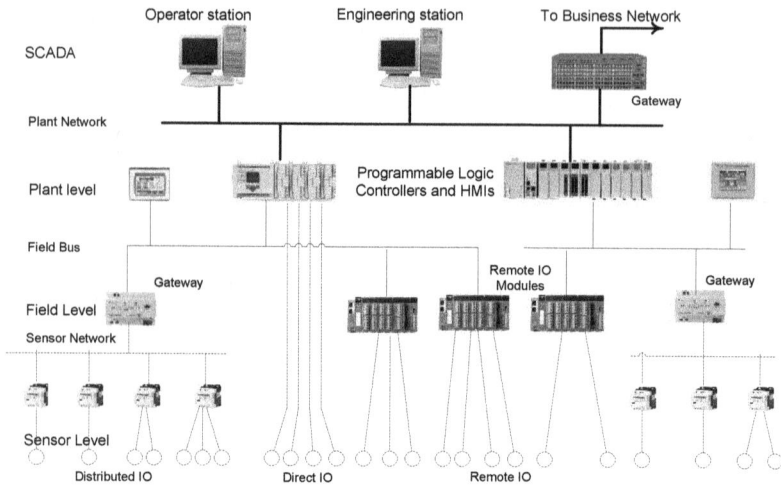

Figure 2.3: Industrial network architecture

## 2.2    Open System Interconnection Reference Model

There are many industrial network protocols with wide variety of capabilities. To appreciate the differences among these protocols, one has to understand how they are implemented with respect to the International Standards Organization's (ISO's) Open System Interconnection (OSI) communication reference model. The model has seven layers, each representing a set of tasks that need to be carried out in order to achieve communication among electronic devices. These layers are fairly independent such that the implementation of tasks in one layer does not hinder or require the implantation of tasks in another layer. Figure 2.4 shows that the OSI reference model is made up of the following layers: application, representation, session, transport, network, data link and physical.

**Application Layer:** This is layer 7 of the OSI reference model. It is responsible for services that directly support user applications, such as software for file transfers, database access, and e-mail. In other words, it serves as a window through which application processes can access network services. This is the point of entry into the OSI reference model for a message being sent across a network, and a point of exist for a received message.

[18]

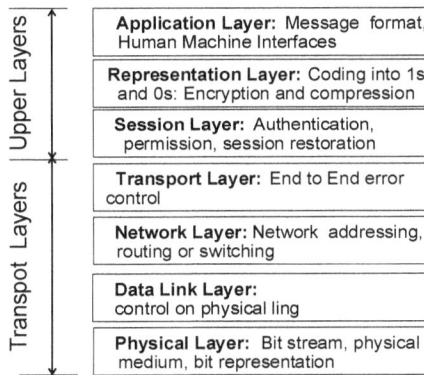

**Figure 2.4: OSI Communication Reference Model**

**Presentation Layer: This is layer 6 of the OSI reference model. It** defines the format used to exchange data among electronic devices. One can think of it as the network's translator. When dissimilar electronic devices such as window and Apple PCs have to communicate, a certain amount of translation and byte reordering must be done. Within the sending device, the presentation layer translates data from the format sent down from the application layer into a commonly recognized, intermediary format; and at the receiving device, this layer translates the intermediary format into a format that can be useful to the receiving device's application layer. The presentation layer is responsible for converting protocols, translating data, encrypting data, changing or converting character sets, and expanding graphics commands. It also manages data compression to reduce the number of bits that need to be transmitted.

**Session Layer**: This is Layer 5 of the OSI reference model. It allows two applications on different electronic devices to open, use, and close a connection called a *session*. It carries out name-recognition and other functions, such as security, that are needed to allow two applications to communicate over the network. The session layer synchronizes user tasks by placing checkpoints in the data stream. The checkpoints break the data into smaller groups for error detection. This way, if the network fails, only the data after the last checkpoint has to be retransmitted. This layer also implements dialog control between communicating processes, such as regulating which side transmits, when, and for how long.

[19]

**Transport Layer:** This is layer 4 of the OSI reference model. It provides an additional connection level beneath the session layer. The transport layer ensures that packets are delivered error free, in sequence, and without losses or duplications. At the sending device, this layer repackages messages, dividing long messages into several packets and collecting small packets together in one package. This process ensures that packets are transmitted efficiently over the network. At the receiving device, the transport layer opens the packets, reassembles the original messages, and, typically, sends an acknowledgment that the message was received. If a duplicate packet arrives, this layer will recognize the duplicate and discard it. Furthermore, the transport layer provides flow control and error handling, and participates in solving problems concerned with the transmission and reception of packets.

**Network Layer:** This is layer 3 of the OSI reference model. It is responsible for addressing messages and translating logical addresses and names into physical addresses. This layer also determines the route from the source to the destination. It determines which path the data should take based on network conditions, priority of service, and other factors. It also manages traffic problems on the network, such as switching and routing of packets and controlling the congestion of data. If the network adapter on the router cannot transmit a data chunk as large as the source computer sends, the network layer on the router compensates by breaking the data into smaller units. At the destination end, the network layer reassembles the data.

**Data-Link Layer:** This is layer 2 of the OSI reference model. The layer sends data frames from the network layer to the physical layer. It controls the electrical impulses that enter and leave the network cable. On the receiving end, the data-link layer packages raw bits from the physical layer into data frames. The data-link layer is responsible for providing error-free transfer of frames from one device to another through the physical layer. This allows the network layer to anticipate virtually error-free transmission over the network connection. Usually, when the data-link layer sends a frame, it waits for an acknowledgment from the recipient. The recipient data-link layer detects any problems with the frame that might have occurred during transmission. Frames that were damaged during transmission or were not acknowledged are then re-sent.

**Physical Layer:** This is layer 1 of the OSI reference model. This layer transmits the unstructured, raw bit stream over a physical medium (such as the network cable). The physical layer is totally hardware-oriented and deals with all aspects of establishing and maintaining a physical link between communicating electronic devices. The physical layer also carries the signals that transmit data generated by each of the higher layers. This layer defines how the cables are attached to the NIC. For example, it defines how many pins the connector has and the function of each. It also defines which transmission technique will be used to send data over the network cable. The layer provides data encoding and bit synchronization. The physical layer is responsible for transmitting bits (zeros and ones) from one computer to another, ensuring that when a transmitting host sends a 1 bit, it is received as a 1 bit, not a 0 bit. Because different types of media physically transmit bits (light or electrical signals) differently, the physical layer also defines the duration of each impulse and how each bit is translated into the appropriate electrical or optical impulse for the network cable.

The remainder of this chapter presents the most common legacy industrial network protocols that implement the OSI reference model requirements to varying level, resulting in widely varying communication capabilities among the protocols. These capabilities determine whether the network is appropriate for use at the sensor, control or SCADA level of the automation system. In fact, as we describe legacy network protocols in this book, we highlight the features which make them appropriate for their applications.

## 2.3    AS-i Network

AS-i stands for Actuator Sensor *Interface. It* is an "OPEN" industrial communication network for data exchange between electro-mechanical input/output devices and automation controllers (mainly PLCs, PCs). AS-i is regulated by the EN 50295/IEC 62026-2 standard, and 10 million nodes of this network have been installed in plants and systems since 1994. One of the key features of AS-i is its ability to carry data and power on the same two-wire cable. This makes it easy and cheap to install. However, it requires a special power supply for the devices. Furthermore, AS-i is intrinsically safe, meaning that its power supply cannot produces sparks, making the network usable in explosive environments without any special modifications. According to Figure 2.2, AS-i is designed for simple logic control, involving simple automation devices such as pilot lamps, push buttons and rotary switch.

### 2.3.1 AS-i Network and the OSI Reference Model

AS-i network implement only three of the seven layers of the OSI reference model, namely: application, data link, and physical. Therefore, the network does not offer communication functions associated with the presentation, session, transport, and network layers of the model and it implements the application, data link and physical layers as follows:

- Application layer (Layer 7): The layer provides the device profiles (device type, e.g. LED, relay, or switch; data tags; and other information about the device). The layer also implements the AS-Interface message as function processes in masters and slave. AS-i network supports 4-bit messages. Therefore, 8 bit analog messages are sent in two cycles and 16 bit messages are sent in five cycles. AS-I network digital messages have a cycle time of less than 5ms while the analog messages have cycle time of less than 10ms.

- Data Link Layer (Layer 2): Supports the Master-Slave (with polling) AS-Interface bus access procedure, structures the messages, and provides the data safeguarding. AS-i has a simple addressing scheme, and the network can handle up to 31 (62 Ver 2.1 (A-B mode)) slaves, where each device can exchange 4 bits of input and 4 bits of output data, resulting in a total of 124 inputs and 124 outputs on a single network with a gross transmission rate of 167 kbit/s.The network update time is easily calculated by multiplying the number I/O nodes with the deterministic update time for each node (approximately 150 microseconds), for a maximum update time of 5ms.

- Physical Layer (Layer 1): AS-i uses a special power supply and a 2-wire unshielded cable that carries both power and data. The maximum cable length on the 2.0 version of the network is 100m (328ft), while that of the 2.1 version is 300 m (984ft). AS-i extension plugs double network length. An AS-i network can have up to three segments with extension plugs, hence increasing the maximum length of the 2.0 version to 600m (1,968 feet). Voltage levels on the network range between 29.5 and 31.6 VDC. In addition to the framing bits, data protection is accomplished via

[22]

Manchester-II coding, a highly symmetrical, floating layout with Alternating Pulse Modulation.

## 2.3.2   AS-i Network Messages

There are two types of AS-i messages: the master requests and the slave responses. Master messages are 14 bit times in length and slave messages are 7 bit times. A bit time corresponds to a uniform 6μs and a master pause is 2 to10 bit times. A master poll of a standard slave consists of the following bits:

- Start Bit (ST). Identifies the beginning of the master request. It is always "0".
- Control Bit (SB). Identifies the type of request: "0" for data, parameter request or address assignment; and "1" for command request such as reset the slave or send ID code.
- Address (A4..A0). Address of the contacted slave address requested (5 bit).
- Information (D4, D3…D0). These 5 bits contains the type of request, and the information to be transferred to the slave.
- Parity Bit (PB). The number of all "1" in the master call has to be even.
- End Bit (EB). Identifies the end of the master request. Always has value "1".

On the other hand, the slave response consists of the following bits:

- Start Bit (SB). Identifies the beginning of the slave response. Its value is always "0".
- Information (D3..D0). These 4 bits represent the properly information sent to the master.
- Parity Bit (PB). The number of all "1" in the slave response has to be even.
- End Bit (EB). Always with value "1", it signals the end of the slave response.

## 2.3.3   Main features of AS-i Network

AS-i networks are usually made up of the following components:

- Embedded Components: I/O devices that connect directly to the AS-i network bus cable.

[23]

- I/O Modules: small "blocks" of general-purpose I/O points, which allow standard I/O devices to communicate through the component network.
- PC Interface Cards and PLC Modules: PLC interfaces and network gateways, which allow host controllers such as PLCs or PCs to control and monitor AS-i network devices.
- Bus Cables and Interconnect Products: cables, connectors, and a complete range of tools and options.
- Set-Up Tools and Configuration Software: hand-held tools and software used to design, document, configure, and commission AS-i network systems and devices.

An AS-Interface network is a collection of network segments, and there are very few rules that need to be satisfied when designing an AS-Interface network:

- There can be no duplicate addresses on a network
- Each segment must be 100m or less in total cable length unless a tuner is used, in which case the segment cable length must not exceed 200m
- Each network must have exactly one master
- Each segment must power exactly one AS-Interface power supply
- When repeaters are used, a slave cannot be more than "two repeater transitions" from the master
- AS-i network can be wired in all sorts of physical topologies, including bus, start, branch and tree topologies. That is, the shape (i.e. topology) of AS-i segments is arbitrary (unrestricted).

Using these basic rules, a linear network, with the master at one end of the network, can be 300m long; while a linear network, with the master "in the middle" of the network, can be 500m In some applications longer networks are desirable (e.g. Process Automation applications). This is possible through the installation of extension plugs.

### 2.3.4    Advantages and Disadvantages of AS-I Network

By using an AS-I network to connect ordinary control and sensing devices, users can save up to 35% of the material and labour cost required to install and commission a control/automation system. Generally, the network has the following advantages over conventional wiring of sensors and actuators:

- Reduced installation cost
- Reduced wiring errors.
- Reduced Troubleshooting Time.
- Increased system uptime.
- Modularity, flexibility, and expandability.
- AS-I networks are complimentary to typical higher-level networks.

AS-I network is deterministic, fast, and intrinsically safe, making it a desired choice for safety networks that allows the connection of safety related devices such as emergence stop, guard switches, light curtains, laser scanners, and gate switches to an intelligent programmable safety relay called a safety monitor over a network. Figure 2.5 shows an AS-I with safety monitor that controls the safety devices on the network. The standard PLC can monitor the state of the safety devices, but it cannot be used to control them.

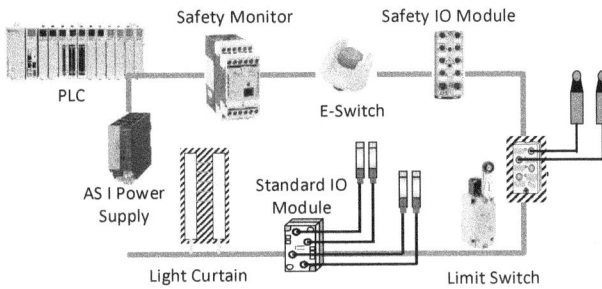

**Figure 2.5: AS-I Network**

The main disadvantage of AS-I network is its inability to carry large pieces of data (limited to 4 bit messages compared to 1520 bytes of TCP/IP). Therefore its applications are limited to the sensor level of the industrial networks or the field level of automation systems.

## 2.4    Modbus

Modbus network protocol was created by MODICON to connect PLCs to programming tools. But it is now widely used to establish master-slave communication between intelligent devices. Modbus is independent of the

[25]

physical layer. It can be implemented using RS232 (EIA232), RS422, RS485 or over a variety of media (e.g. fiber, radio, cellular, Ethernet etc). It is managed by a Telecommunications Industry Association / Electronic Industries Alliance (TIA/EIA). According to Figure 2.2, Modbus is a lower midlevel network. It usually used to integrated small controllers such as small PLCs and microcontrollers with large complex controls such as advance PLCs and DCS systems. Some flavors of Modbus are used at the plant level of the IEC automation hierarchy.

### 2.4.1    Modbus Network and the OSI Reference Model

Modbus has three main flavors, namely: Modbus Serial, Modbus Plus and Modbus TCP. Figure 2.6 shows that these flavors are implemented differently according to the OSI references model; resulting into the following capability differences:

- Modbus Serial: Uses RS485 as the main communication medium, and master/slave media access control method. Its transmission speed ranges from 1,200bits/s to 115 Kbits/s. The Modbus serial comes in 2 versions: ASCII transmission mode, in which each eight-bit byte in a message is sent as 2 ASCII characters and RTU transmission mode for which each eight-bit byte in a message is sent as two four-bit hexadecimal characters. The main advantage of the RTU mode is that it achieves higher throughput, but ASCII mode allows time intervals of up to 1 second to occur between characters without causing an error.
- Modbus TCP/IP: It uses TCP/IP and Ethernet 10 Mbit/s or 100 Mbits/s to carry the Modbus messaging structure. This protocol is covered in detail in Section 4.4.
- Modbus Plus: This is a higher speed network (1 Mbit/s) token passing derivative that uses the Modbus messaging structure and RS48 communication medium. It is a proprietary specification of Schneider Electric; and instead of being patented, it is unpublished. Normally, the protocol is implemented using a custom chipset available only to partners of Schneider.

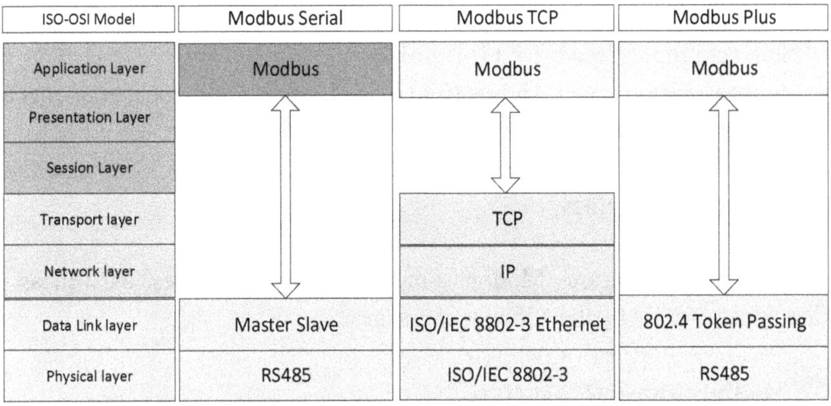

| ISO-OSI Model | Modbus Serial | Modbus TCP | Modbus Plus |
|---|---|---|---|
| Application Layer | Modbus | Modbus | Modbus |
| Presentation Layer | | | |
| Session Layer | | | |
| Transport layer | | TCP | |
| Network layer | | IP | |
| Data Link layer | Master Slave | ISO/IEC 8802-3 Ethernet | 802.4 Token Passing |
| Physical layer | RS485 | ISO/IEC 8802-3 | RS485 |

Figure 2.6: Flavors of Modbus Protocol

## 2.4.2 Modbus Data and Function Structure

The reference to Modbus data location or entities (coils, discrete inputs, input registers, holding registers) is governed by some conventions. In the traditional standard, *numbers* for those entities start with a digit, followed by a number of four digits in range 1 - 9,999 as follows:

- Coils *numbers* start with a **zero** and then span from **0**0001 to **0**9999.
- Discrete input *numbers* start with a **one** and then span from **1**0001 to **1**9999.
- Input register *numbers* start with a **three** and then span from **3**0001 to **3**9999.
- Holding register *numbers* start with a **four** and then span from **4**0001 to **4**9999.

This limits the number of *addresses* to 9,999 for each entity. The general referencing scheme extends this to the maximum of 65,536, by one digit to the previous list as follows:

- Coil *numbers* span from **0**00001 to **0**65536.
- Discrete input *numbers* span from **1**00001 to **1**65536.
- Input register *numbers* span from **3**00001 to **3**65536.
- Holding register *numbers* span from **4**00001 to **4**65536.

The various functions (services) supported by Modbus applications can be categorized as follows:

[27]

- Coils: read and write, 1 bit (off/on)
- Discrete Inputs: read, 1 bit (off/on)
- Input Registers: read, 16 bits (0 to 65,535), essentially measurements and statuses
- Holding Registers: read and write, 16 bits (0 to 65,535), essentially configuration values

Table 1 presents the various Modbus functions, their categories, as well as the function codes carried by the Modbus messages.

### 2.4.3    Modbus Message Structure

The Modbus frame structure is the same for requests (master to slave messages) and responses (slave to master messages). Figure 2.7 shows the frame structure for Modbus RTU and Modbus ASCII. The frames have the following main components:

**Address Field:** The slave devices are assigned addresses in the range of 1 ... 247. Address 0 is reserved for broadcast messages (no response). Request: A master addresses a slave by placing the slave address in the address field of the message. Response: When the slave sends its response, it places its own address in this address field of the response to let the master know which slave is responding.

**Function Field:** Valid codes are in the range of 1 ... 255 decimal. Request: The function code field tells the slave what kind of action to perform. Response: For a normal response, the slave simply echoes the original function code. For an exception response, the slave returns a code that is equivalent to the original function code with its most significant bit set to a logic 1.

**Data Field:** Request: The data field contains additional information which the slave must use to take the action defined by the function code. This can include items like register addresses, quantity of items to be handled, etc. Response: If no error occurs, the data field contains the data requested. If an error occurs, the field contains an exception code that the master application can use to determine the next action to be taken.

[28]

#### Table 2.1: Modbus Functions Codes

| Function type | | | Function name | Code |
|---|---|---|---|---|
| Data Access | Bit access | Physical Discrete Inputs | Read Discrete Inputs | 2 |
| | | Internal Bits or Physical Coils | Read Coils | 1 |
| | | | Write Single Coil | 5 |
| | | | Write Multiple Coils | 15 |
| | 16-bit access | Physical Input Registers | Read Input Registers | 4 |
| | | Internal Registers or Physical Output Registers | Read Multiple Holding Registers | 3 |
| | | | Write Single Holding Register | 6 |
| | | | Write Multiple Holding Registers | 16 |
| | | | Read/Write Multiple Registers | 23 |
| | | | Mask Write Register | 22 |
| | | | Read FIFO Queue | 24 |
| | File Record Access | File Record Access | Read File Record | 20 |
| | | | Write File Record | 21 |
| | Diagnostics | | Read Exception Status | 7 |
| | | | Diagnostic | 8 |
| | | | Get Com Event Counter | 11 |
| | | | Get Com Event Log | 12 |
| | | | Report Slave ID | 17 |
| | | | Read Device Identification | 43 |
| Other | | | Encapsulated Interface Transport | 43 |

**Error Check Field:** Modbus carries out parity check (even/odd) on every word. The parity bits are carried in the fields associated with the checked words. The Error Check field carries the frame checking data which is normally LRC (Longitudinal Redundancy Check) or CRC (cyclic redundancy check).that is applied to the entire message. In addition to word and frame checking, Modbus identifies communication errors using the continuous stream method. Under this method, the entire message frame must be transmitted as a continuous stream. If a silent interval (more than 1.5 character times RTU mode or 1 second ASCII mode) occurs before completion of the frame, the receiving device flushes the incomplete message and assumes that the next byte will be the address field of a new message.

### 2.4.4    Main Features of Modbus

Modbus network can carry more data than AS-i network, therefore it can be used at the field level of the automation system where the amount of the data that need to be transferred is higher than at the sensor level. On the other hand Modbus Serial is based on three layers of the OSI like AS-i making it deterministic and generally

[29]

fast. Therefore, the network is very common in advance sensor and actuator applications such as bar code readers, printers and scales. Generally, Modbus has the following main features:

- Topology: Bus with line terminations
- Maximum distance: With RS485 : 1000 m without repeater
- Data rate: From 1,200 to 115 Kbits/s
- Max. no. of devices:     With RS485 : 32 master included
- Method of accessing the medium: Master slave
- Transmission method: Messaging
- Max. useful data size: 120 words
- Transmission security: LRC or CRC, start and stop delimiters, parity bit and continuous stream

Modbus RTU

| silence | Address | Function | Data | | CRC | | silence |
|---|---|---|---|---|---|---|---|

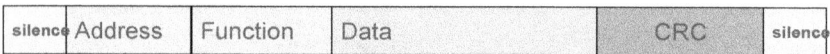

Silence >= 3,5 characters

Modbus ASCII

| : | Address | Function | Data | | CRC | | CR | LF |
|---|---|---|---|---|---|---|---|---|

3A Hex

0D Hex
Carriage return

0A Hex
Line Feed

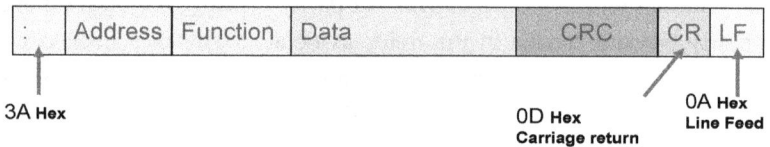

**Figure 2.7: Modbus Message Frame**

## 2.5    DeviceNet

DeviceNet is a fieldbus that provides connections between sensors, actuators and PLCs and PCs. It has master/slave and peer to peer capabilities. It is managed by Open DeviceNet Vendor Association (ODVA) and it is used with drives, bar-code RFID, and remote I/O (DI/DO, ADC/DAC). DeviceNet is an open industrial network built upon the Common Industrial Protocol - (CIP), the same upper layers protocol up on which Ethernet IP and ControlNet are based (Figure 2.8). According to Figure 2.2, DeviceNet is a midlevel network. This means that it should be used at the control level of the IEC automation hierarchy. While it is used as such, DeviceNet also finds a lot of applications at the sensor level where it is used to connect simple automation devices such as limit switches, stacked lamps, and other field devices to PLCs.

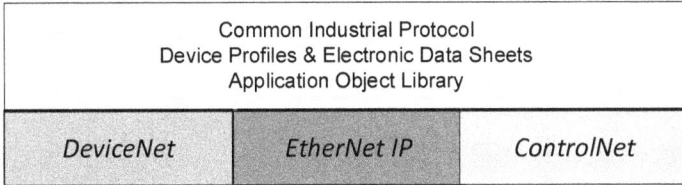

| Common Industrial Protocol<br>Device Profiles & Electronic Data Sheets<br>Application Object Library | | |
|:---:|:---:|:---:|
| *DeviceNet* | *EtherNet IP* | *ControlNet* |

**Figure 2.8: The Control and Information Protocol**

## 2.5.1   DeviceNet and the OSI Reference Model

Figure 2.9 shows that DeviceNet implements all the seven layers of the OSI reference model. The upper three layers are implemented by the Communications and Information Protocol (CIP), formerly the Common Industrial Protocol (CIP) which defines the device profiles, data structure, as well as the communication services provided by DeviceNet. DeviceNet implements the other layers of the OSI model as follows:

**Transport Layer:** This layer is implemented according to IEC62026 (DeviceNet Specification) standard. It is responsible for the fragmentation of data if more than 8 bytes need to be sent. In such cases, only 7 bytes of data can be sent in the frame as the first byte is used to indicate fragment numbers.

**Network Layer:** This layer is implemented as a single unit with the transport layer based on IEC62026 standard. It supports DeviceNet's feature of being a connection-based network.  A connection is initially established by an UCMM (Unconnected Message Manager), before explicit and implicit messages are sent or received.

**Data Link layer:** This layer implements the Media Access Control (MAC) method. DeviceNet uses two MACs, namely master-slave and CSMA (CA-NBA), hence the ability to support advanced messaging schemes such as implicit producer-consumer (peer-to-peer) and change-of-value. The device addresses are used in the arbitration processes, and in case of data collision devices with lower addresses continue to communicate. Once a master has won the arbitration, it communicates with its slaves using master-slave MAC.

[31]

DeviceNet frames are structured as follows:

- Start of Frame (1 byte)
- Identifier, Remote Transmission Request(11 bits,1 byte)
- Control(6 bits)
- Data( 0 to 8 Bytes)
- CRC sequence, CRC Delimiter(15 bits,1 bit)
- ACK slot, ACK Delimiter(1 bit,1 bit)
- End of Frame(7 bits)
- Interframe Space ( 3 or >3 bits)
- Delimiter is used to report to sender that message was not received

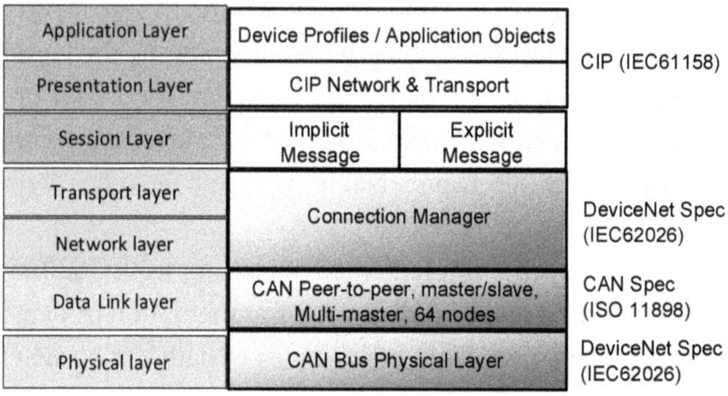

| | | |
|---|---|---|
| Application Layer | Device Profiles / Application Objects | CIP (IEC61158) |
| Presentation Layer | CIP Network & Transport | |
| Session Layer | Implicit Message | Explicit Message |
| Transport layer | Connection Manager | DeviceNet Spec (IEC62026) |
| Network layer | | |
| Data Link layer | CAN Peer-to-peer, master/slave, Multi-master, 64 nodes | CAN Spec (ISO 11898) |
| Physical layer | CAN Bus Physical Layer | DeviceNet Spec (IEC62026) |

**Figure 2.9: DeviceNet and the OSI Communication Reference Model**

**Physical layer:** The physical topology is a bus network. The bus or trunk is terminated at both ends by 120 ohm ¼ W resistors. The network uses three types of five-wire cables: thick, thin, and flat wire, as well as 5 pin connectors. One wire is the drain, two of the wires carry the signal, and the other two carry power.

DeviceNet uses the following signaling:

- Two wire differential network, CAN_H swings between 2.5 V( recessive state) and 4.0 V DC( dominant state) , while CAN_L swings between 2.5 V( recessive state) and 1.5 V DC( dominant state) relative to V- ( power supply ground)
- With no network master connected, the CAN_H and CAN_L lines should be between 2.5 V and 3.0 V

- With a network master connected and polling the network, the CAN_H should be around 3.2 V and CAN_L 2.4 V. This is because the signals are switching and this affects the DC V read by the meter
- The nodes are attached to the bus in parallel resulting in a wired-AND configuration meaning only when all nodes output a high signal, the signal on the bus is high too

## 2.5.2    DeviceNet Messaging

The DeviceNet protocol supports two basic kinds of message - cyclic I/O and explicit message. Each message types is associated with a particular type of data, as described below:

Cyclic I/O: This is a synchronous telegram type of messaging used for processing priority data between a producer and one or more consumers. They are divided according to the data exchange method. The main ones are:

- Polled: This is method of communication where the master sends a telegram to each one on its slave list (scan list). As soon as the request is received, the slave responds promptly to the master. This process is repeated until everyone is consulted and the cycle restarts.
- Bit-strobe: method of communication where the master sends the network a telegram with 8 data bytes. Each bit of these bytes represents a slave that responds according to its program, if addressed.
- Change of State: method of communication where the exchange of data between master and slave only occurs if the values monitored/controlled change, up to a time limit. When this limit is reached, the transmission and the reception will go on, even without alterations. The configuration of this time variable is executed in the network configuration program.
- Cyclic: another method of communication similar to the previous one. The only difference is in the message production and consumption.   In this type, every data exchange is performed in regular time periods regardless of being altered or not. These intervals are also adjusted in the network configuration software.

*Explicit Message*: type of non-priority, general use telegram. Used mainly in non-synchronous jobs like parameterization and configuration of equipments.

### 2.5.3 Key Features of DeviceNet

DeviceNet has the following key features:

- Based on ISO 11898 (CAN) –CSMA(CA-NBA)
- Every DeviceNet nodes have associated configuration files, called EDS (Electronic Data Sheet). This file keeps important data on the device work and must be registered on the network configuration software.
- Media access arbitration is based on node addresses. Messages from devices with lower addresses have a higher priority,
- Producer/Consumer and Master/Slave
- Trunk line/drop line
- 64 max nodes
- 1 bit to multiple byte size message
- 125 kbps@ 500 m, 250 kbps @ 250 m and  500 kbps @ 100 m
- Any 24 V power supplies can be used
- Max 8 A current limit on thick/flat wire and 3 A on thin wire
- Voltage range at each node 11.0 to 25 volts
- Max voltage drop of 10 V across each node

## 2.6    Profibus

Profibus (**Process Field Bus**) originated in the European market in the late 1980s. It has become a worldwide standard. There are three versions of Profibus, namely: Profibus FMS (Fieldbus Message Specification), Profibus-DP (Decentralized Periphery) and Profibus-PA (Process Automation).

**Profibus FMS** was the initial version of Profibus, and it was designed to send complex information between Programmable Controllers and PCs. This protocol turned out not as flexible as needed for industrial automation, and it is not appropriate for less complex messages or communication on a wider, more complicated network. Profibus FMS is not widely used therefore we do not discuss it any further in the book.

**Profibus DP** is much simpler and faster than Profibus FMS. It was designed to support communication between controllers and remote IOs. It is also used to replace parallel signal transmission with 24 V or 0 to 20 mA. Profibus DP has, itself, three separate versions. Each version, from DP-V0 to DP-V1 and DP-V2, provides newer, more complicated features.

**Profibus PA** was designed for Process Automation. It uses Profibus DP application profile, and it standardizes the process of transmitting measured data. In most environments, Profibus PA operates over RS485 twisted pair media. This media, along with the PA application profile supports power over the bus. In explosive environments, though, that power can lead to sparks that induce explosions. To handle this, PROFIBUS PA is used with Manchester Bus Powered technology (MBP).

### 2.6.1 Profibus and the OSI Reference Model

Like AS-i network, is based on the reduced OSI communication reference model so as to provide increased speed and determinism. But this means that Profibus does not have functions related to the presentation, session, and transport layers of the OSI communication reference model (Figure 2.10).

**Application layer:** At the application layer Profibus FMS has its own specification; while Profibus DP and PA use the same DP-V0, DP-V1 and DP-V2 specifications to define the data structure as well as messaging services (see Figure 2.10). The different versions of Profibus handle different types of messaging at the application layer. Some of the types of messaging supported include cyclic and acyclic data exchange, diagnosis, alarm-handling, and isochronous messaging.

**Data Link Layer:** Profibus implements the data link layer through a Fieldbus Data Link, or FDL. The FDL system combines two common schemes, master-slave methodology and token passing. Therefore, Profibus is a multi-master, multi-slave network. The token is passed among the masters, and only the master with the token can communicate with its slaves, using the master-slave paradigm.

**Physical Layer:** Profibus supports three types of physical media. The first is a standard twisted-pair wiring system (RS485), and the second is a more advanced systems that uses fiber-optic transmission. The third physical medium is reserved for Profibus PA, It is a safety-enhanced system called Manchester Bus Power, or MBP. This medium makes Profinet PA intrinsically safe and useable in situations where the chemical environment is prone to explosion.

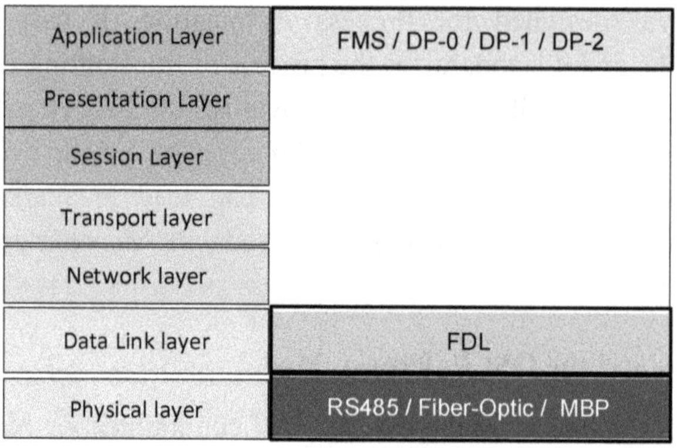

| Application Layer | FMS / DP-0 / DP-1 / DP-2 |
|---|---|
| Presentation Layer | |
| Session Layer | |
| Transport layer | |
| Network layer | |
| Data Link layer | FDL |
| Physical layer | RS485 / Fiber-Optic / MBP |

**Figure 2.10: Profibus and the OSI reference Model**

## 2.6.2    Profibus DP

Profibus DP is one of the most widely used fieldbuses in industry. It is designed such that each station is given a unique address which should be a number between 1 and 125. In addition, Profinet DP has the following characteristics when using RS485 communication medium:

- Maximum cable length is (100m at 12 Mbit/s, 1200m at 9.6 kBit/s).
- If the cables are long or the number of stations exceeds 32 (master included), there is a need for repeaters.
- The total length of the network cannot exceed 10 km.

Profibus DP can operate using fiber-optic transmission in cases where that is more appropriate. This enables the network to have the following features:

- Noise immunity.
- Potential difference independence.
- Longer distances (up to 20 miles).
- Redundant operation capability.
- Line, ring and star configuration.

**Profibus DP Network Configuration:** Standardized electronic data sheets called General Station Description (GSD) files are used to permit open, vendor-independent configuration. Each Profibus device has a GSD; and a library with GSD files of Profibus devices is available online at www.PROFIBUS.com. The files are uploaded into controller IDEs, giving the IDEs access to the devices

[36]

description, data structure, and application services. After configuring the controller through the IDE, the GSD files are downloaded into the controller together with the process control logic.

### 2.6.3    ProfiBus PA

Profibus PA is a Network of field devices that communicate with Profibus DP controllers. It is used in process automation to replace HART Protocol. PROFIBUS PA meets 'Intrinsically Safe' (IS) and bus-powered requirements defined by IEC 61158-2, the fieldbus standard used throughout process automation. A PROFIBUS PA fieldbus network can connect up to 32 devices per 'segment' depending on the type of devices and the application. Furthermore, Probus PA supports a baud rate of 31.25kbits/s.

### 2.6.4    Profibus PA/Profibus DP Interfaces

There are two main types of PD/PA interfaces, namely: Transparent and Non-Transparent interfaces.

- Transparent PD/PA interfaces use couplers: For these interfaces, the controller sees the PA devices directly (by address) as its slaves, and the device addresses are limited to 1-125.
- Non Transparent PD/PA interfaces use links and couplers: The controller does not see the PA devices directly by address, and links have Profibus DP addresses. We can have up to 5 couplers per link that are limited to 1-125. Device addresses can be reused among links.

### 2.6.5    Profibus PA Network Setup – Rules

Figure 2.12 show the implementation of Profibus PA as a sensor network for a Profibus DP network. For the Profibus PA to operate correctly, its design must meet the following specifications:

- No more than 5 couplers can be placed on a single link
- Only up to 31 PA field devices can be connected to single link - INDEPENDENT of the number of couplers
- Maximum current flow per coupler is:
- 400mA for none safe system
- 110mA for EEx [ib] II C ignition protection
- 90mA for EEx [ia] II C ignition protection

[37]

- Maximum segment length per coupler = 1,900m
- Spur line max. 30 m , and spur line current must no more than 10mA
- Maximum cable length depends on type of cable used for the network.

## 2.7    References

1.  Cheng-Yuan. Hsieh , Introduction to Computer Networks, accessed from http://pluto.ksi.edu/~cyh/cis370/ebook/ch05b.htm
2.  *Binder, Kai (2012-11-09).* "Einfach – sicher - international". *SPS Magazin.*
3.  Jump up "AS-Interface – Experts Forum". 091119 as-interface.net
2.  Ataide, F.H. (2004). Estudo Técnico EST-DE-0007-04 - AS-Interface, SMAR Equipamentos Industriais Ltda, fevereiro.
3.  Becker, R.; Müller, B.; Schiff, A.; Schinke, T.; and Walker, H. (2002). AS-Interface - The Automation Solution, AS-International Association, Germany.
4.  *Schneider Electric "Modbus Plus - Modbus Plus Network - Products overview - Schneider Electric United States". Schneider-electric.com.* Retrieved 2014-01-03.
5.  Modbus Organization Inc. "Modbus Application Protocol V1.1b3", *Modbus. Modbus Organization, Inc.* Retrieved 2 August 2013.
6.  Clarke, Gordon; Reynders, Deon (2004*).* Practical Modern Scada Protocols: Dnp3, 60870.5 and Related Systems. *Newnes. pp. 47–51.* ISBN 0-7506-5799-5.
7.  Control Solutions, Inc., Jump up "Modbus 101 - Introduction to Modbus", Accessed from: http://www.csimn.com.
8.  Real Time Automation, DeviceNet Unplugged – A View "Under the Hood" for End Users, White paper, Accessed from http://www.rtaautomation.com
9.  SMAR, DeviceNet, Tutorial, Accessed from http://www.smar.com
10. Profibus International Organization, Profibus-Profinet, available at http://www.profibus.com/tec

## 2.8    Discussion Questions

### Question 1

The slave device in Figure 2.11 has 24 V DC sourcing transistor outputs. Use the Figure to answer question 1 of Section B:

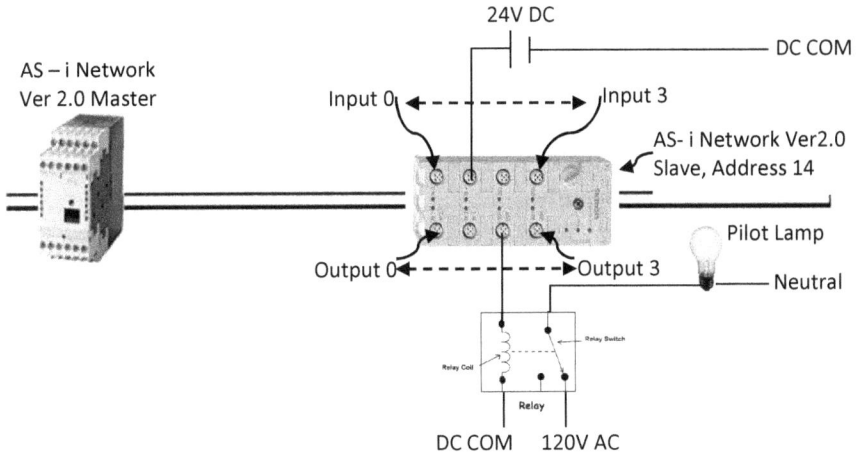

**Figure 2.11: Schematic Diagram of AS-i Network based Pilot Lamp Control System**

i)     What is the message the master sends to the slave to request for the status of slave input terminals?

ii)    What is the response of the slave to the request of the master asking for the status of the slave input terminals?

iii)   What is the request message the master sends to the slave to turn on the pilot lamp?

iv)    What is the request message the master sends to the slave to turn off the pilot lamp?

v)     What is the request message the master sends to the slave to change its address to 10?

### Question 2

i)     Most Common physical medium of Modbus RTU is ...................

ii)    Modbus RTU accommodates........ Masters.

iii)   The number of slaves on a Modbus network is determined by..............

iv)    Modbus RTU master request to read holding registers is ...........bytes long.

[39]

v)    What is the difference between Modbus RTU master request message frame and slave response message frame?

**Question 3**

i)    What is the minimum number of PROFIBUS PA links are required to connect 55 PROFUIBUS PA devices to a PROFUIBUS DP network?

ii)    Similar 61 PROFIBUS PA devices need to be connected to a network, and each requires a current of 105mA. In addition, imagine that the network is not meant for a safe system. What is the minimum number of Links required to connect the PROFIBUS PA devices to a PRIFIBUS DP network.

iii)    If the devices have ignition protection rated EEx [ib] II C and the network is required to be safe, What is the minimum number of Links required to connect the PROFIBUS PA devices to a PRIFIBUS DP network.

# Chapter 3  *Ethernet and TCP/IP in Industrial Networks*

In order to create new opportunities, increase efficiency of existing plants and processes, and reduce operating costs, manufacturing companies are increasingly expanding their global operations. But to achieve their globalization goals and operational excellence requires improved connectivity between plant and business systems for real-time visibility of information and effective collaboration, this in turns leads to the following benefits:

- consistent quality of products and performance across global operations
- Reduced cost of design, deployment, and support of distributed manufacturing and IT systems.
- Improved balance between production and demand to optimize material and asset utilization.
- Improved response to events that occur on the plant floor, regardless of location, while implementing more flexible and agile operations in order to react to rapidly changing market conditions.

Advancements in industrial Ethernet based communication protocols are making it possible to use Ethernet technology in real-time systems. Therefore, as manufacturers seek to improve processes, increase productivity, reduce operating costs, and integrate manufacturing and business networks, they are turning to using Ethernet technology on the factory floor.

## 3.1    Industrial Ethernet and the OSI Reference Model

Figure 3.1 shows the relationship between the OSI communication reference model and the TCP/IP stack. Moreover, the figure shows the following:

- The Ethernet layer of the TCP/IP stack is made up of the physical and data link layers of the OSI model.
- The Internet layer of the TCP/IP stack corresponds to the network layer of the OSI model.
- The upper layer of the TCP/IP stack is implemented as a single application layer.

[41]

- The transport, Internet, and Ethernet layers of the TCP/IP stack make up the T-profile of the OSI model.

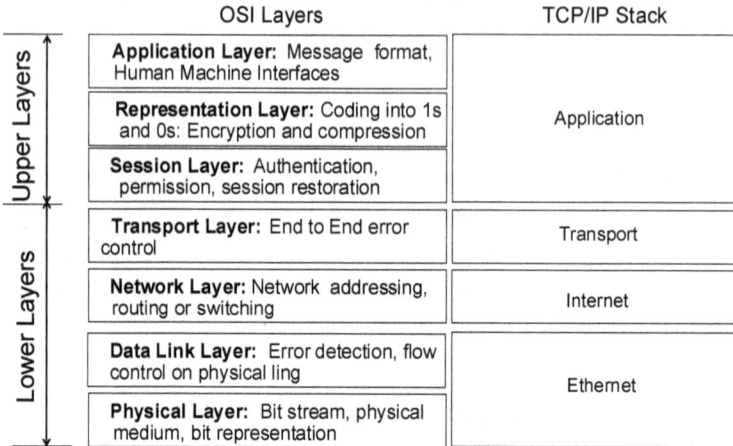

| OSI Layers | | TCP/IP Stack |
|---|---|---|
| Upper Layers | **Application Layer:** Message format, Human Machine Interfaces | Application |
| | **Representation Layer:** Coding into 1s and 0s: Encryption and compression | |
| | **Session Layer:** Authentication, permission, session restoration | |
| Lower Layers | **Transport Layer:** End to End error control | Transport |
| | **Network Layer:** Network addressing, routing or switching | Internet |
| | **Data Link Layer:** Error detection, flow control on physical ling | Ethernet |
| | **Physical Layer:** Bit stream, physical medium, bit representation | |

**Figure 3.1: OSI Layers and TCP/IP Stack**

Generally, Ethernet based communication protocols implement the following layers of the OSI model:

**The application layer**: Ethernet based communication protocols implement the upper three layers as a single application (application that depends on a particular communication protocol) layer. An application is any set of tasks that are carried out above the transport layer. This includes all of the processes that involve user interaction. The application determines the presentation of the data and controls the session. In TCP/IP the terms **socket** and **port** are used to describe the path over which applications communicate. There are many application level protocols in TCP/IP, for example, Simple Mail Transfer Protocol (SMTP) and Post Office Protocol (POP) used for e-mail, Hyper Text Transfer Protocol (HTTP) used for the World-Wide-Web, File Transfer Protocol (FTP) and Trivial File Transfer Protocol (TFTP) Others in this category are: DNS and TELNET Most application level protocols are associated with one or more port number

**The transport layer:** This is supported by two protocols, namely Transmission Control Protocol (TCP) and User Datagram Protocol (UDP). TCP guarantees that information is received as it was sent. On the other hand UDP does not perform end-to-end reliability checks. Some application layer protocol use TCP (e.g. FTP, SMTP, TELNET) and others use UDP (TFTP, SNMP, BOOTP)

[42]

**The Internet layer:** In the OSI reference model the network layer isolates the upper layer protocols from the details of the underlying network that manages the connections across the network. The Internet Protocol (IP) manages the TCP/IP network layer. Because of the inter-networking emphasis of TCP/IP, the IP layer is commonly referred to as the Internet layer. Other protocols that can be implemented in this layer are: ARP, RARP, and Internet Control Message Protocol (ICMP)

**The Ethernet layer:** In TCP/IP, the data link layer and physical layer are normally combined together into the Ethernet layer. The Ethernet layer makes use of existing data link and physical layer standards rather than defining its own. The datalink layer is responsible for Media Access Control (MAC) and error detection. Ethernet uses Carrier Sense Multiple Access with Collision Detection (CSMA-CD) for the MAC, and Cyclic Redundancy Check (CRC) for communication error detection. The physical layer typically defines the characteristics of the hardware that carries the communication signal, such as pin configurations, voltage levels, and cable requirements. Ethernet physical medium can be coaxial cable, twisted pair, optical fiber or wireless.

During transmission, an Ethernet frame starts off as data from the application layer. This data may have an application layer protocol header, such as the HTTP header for a web application. When the data is processed at the different layers, the layer headers are added to it (Figure 3.2). The reverse happens on reception. When the frame is received at the physical layer, processing at the various layers remove the associated header until data (payload) is delivered to the associated application.

Simple Ethernet Frame

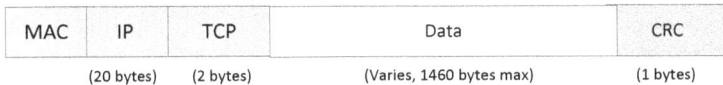

| MAC | IP | TCP | Data | CRC |
|-----|-----|-----|------|-----|
|  | (20 bytes) | (2 bytes) | (Varies, 1460 bytes max) | (1 bytes) |

**Figure 3.2: Ethernet Frame**

The maximum length of data that can be transmitted by a protocol in one instance is called the Maximum Transmission Unit (MTU). For Ethernet, the MTU is 1500 bytes by default. This excludes the Ethernet frame header and trailer. In its simplest form, an Ethernet frame has 20 byte IP header plus 20 byte TCP header, leaving a 1460 byte of the payload that can be transmitted in one frame. This is referred to as maximum segment size (MSS) The MAC (Ethernet) header is 14 bytes and the

[43]

CRC is 4 bytes. More headers may be added to identify the frame, VLAN frame, VPN frame, and so on.

Designing and implementing Ethernet based industrial networks requires two skill sets. The first set of skills is associated with the lower layers, less the transport layer of the OSI reference model. It involves designing the physical network, and understanding various network principles such as addressing structure, wiring requirements, and node power requirements. This skill set is general to all Ethernet based networks. The second skill set is associated with the upper layers (application layer of Ethernet based protocols) plus the transport layer of the OSI model. It involves configuration and programming communications objects in device Integrated Development Environments (IDEs). In this book we present laboratories that cover both skill sets.

## 3.2    Ethernet Networks

This section deals with concepts of Ethernet based networks that are associated with the lower layers of the OSI communication reference model. Most Ethernet based protocol covered in this book identify their hosts (nodes) by IP addresses. Moreover the addressing scheme determines data follow in the networks, something that is very important to industrial networks since they require determinism and speed. IP addresses based on IP Version 4.0, have the following features:

- An IP address consists of 32 bits,

- An IP address is often shown as 4 numbers, each between 0-255 represented in decimal form instead of binary form.

- An example of IP address: 168.212.226.204; in binary form it is 10101000.11010100.11100010.11001100.

- IP addresses are classified into five categories based on the first octet of the address. In fact, IP addresses are made up of two parts. The first part identifies the network, and the second part identifies the host (Figure 3.3).

### 3.2.1    Subnet Mask and Number of hosts

Subnet mask is a 32 bit address used with an IP address to identify network and host portions of the address (Figure 3.3). For example, IP address 200.1.1.2 with a subnet mask 255.255.255.0, 200.1.1 is the network portion and last octet is the host portion. This network can have a maximum of ($2^8$-2 = 254) hosts, because the lowest IP address in the subnetwork, 200.1.1.0 is used to identify the network, while the highest IP address in the subnetwork 200.1.1.255 is reserved for broadcast.

|  | 0 | 1 | 2 | 8 | 16 | 24 | 31 |
|---|---|---|---|---|---|---|---|
| Class A | 0 | Network Id | | | Host Id | | |
| Class B | 1 | 0 | | Network Id | | Host Id | |
| Class C | 1 | 1 | 0 | | Network Id | | Host Id |
| Class D | 1 | 1 | 1 | 0 | Multicast Address | | |
| Class E | 1 | 1 | 1 | 1 | 0 | Reserved for future use | |

**Figure 3.3: The Five Types of IP Addresses**

## 3.2.2 Subnetworks

Creating subnetwork of a network has the following advantages:

- Simplifies network administration as problems in one subnetwork may be addressed without disrupting the other subnetworks
- Enables use of different physical media such as Ethernet and FDDI.
- Adds another layer of security since subnetworks can be designed such that users in one subnetwork do not access hosts in another subnetwork. This is particularly important in industrial automation where individual machines may be assigned their own subnetwork that one may not want the general network users to access.

Subnetworks are created by borrowing some of the host identification bits and using them as part of the network identification bits. The bits must be taken consecutively starting from left to right. Imagine a company's network address is 192.168.2.0 with a subnet mask of 255.255.255.0. If it is required to create two subnetworks, the information of this company network would be represented in binary and decimal as shown in Table 3.1. The host bit X that is borrowed to be part of the network bits can take on a value of 0 or 1, producing two network addresses. Therefore, Table 3.1 shows that after taking one of the bits from the host identification portion of the IP address, we produce two possible network identifications, namely: 192.168.2.0, and 192.168.1.2.128. This leaves only seven

[45]

bits to identify hosts, leading to a maximum of $(2^7 - 2 = 126)$ hosts per subnetwork, and a total of $(126 \times 2 = 252)$ hosts on the entire network. Note that subnetworking reduces the total number of hosts on the network, which is a tradeoff for the above identified advantages of subnetworking. Generally, if n bits are borrowed out of m bits that are originally used to identify hosts, a total of $2^n$ subnetworks are produced. Each subnetwork can have maximum of $(2^{m-n} - 2)$ hosts, leading to a total of $(2^{m-n} - 2) \times 2^n$ hosts for the entire network.

**Table 3.1: IP Addresses, Subnetworks and Subnet Masks**

| | | | | |
|---|---|---|---|---|
| Original Network Address (Decimal) | 192 | 168 | 2 | 0 |
| Original Subnet Mask(Decimal) | 255 | 255 | 255 | 0 |
| Original Network Address (Binary) | 11000000 | 10101000 | 00000010 | 00000000 (Host Id Portion) |
| Original Subnet Mask (Binary) | 11111111 | 11111111 | 11111111 | 00000000 |
| Network Address - One Bit Taken | 11000000 | 10101000 | 00000001 | X 0000000 (Host Id Portion) |
| Subnet Mask –One Bit taken (Binary) | 11111111 | 11111111 | 11111111 | 1 0000000 |
| Subnet Mask –One Bit taken (Decimal) | 255 | 255 | 255 | 128 |
| 1st Possible Network Address (Binary) | 11000000 | 10101000 | 00000010 | 0 0000000 (Host Id Portion) |
| 1st Possible Network Address (Decimal) | 192 | 168 | 2 | 0 |
| 2nd Possible Network Address (Binary) | 11000000 | 10101000 | 00000010 | 1 0000000 (Host Id Portion) |
| 2nd Possible Network Address (Decimal) | 192 | 168 | 2 | 128 |

Figure 3.4 shows the physical layout of the network described in table 3.1. Note that the hosts on the two subnetworks can communicate with each other. This is a desirable situation in most business networks where the focus of subnetwork is to simplify administration. This network is implemented using a router that has OSI reference model level 3 LAN ports or level 2 ports that have VLAN capability. The port settings are shown in Figure 2.6, and the network works as follows: for host on LAN A, say one with IP address 192.168.2.2 and one on LAN B with IP of 192.168.2.130:

- If the host on LAN B tries to connect to the host on LAN A, the packets go to LAN B's switch, which notices that the host is not on its subnet and it forward the packet up the chain to router.
- Router notices that the destination host is on its network, and on subnetwork 192.168.2.0. It therefore forwards the packets to LAN A's switch.
- LAN A's switch recognizes the destination host as being on its subnetwork, and forwards the packets to it.

In addition, the network in Figure 3.4 works as follows when a host tries to connect to another host located on the internet:

- A host on LAN B tries to connect to the host on the Internet, the packets go to LAN B's switch, which notice that the host is not on its subnet and it forward the packet up the chain to the router.
- The router notices that the destination host is on its network, and forward the packets to the modem, making the host on LAN B able to access the Internet.
- A similar scenario would play out if a host on LAN A tried to access the Internet.

The Network in Figure 3.5 is design such that hosts on either subnetwork do not communicate. This is achieved by setting the two routers up using the same subnet but on different "LAN"s. Moreover the Y network with IP addresses of 192.168.0.X, connecting the routers cannot route packets from one LAN to the other. This network works as follows:

- Host on LAN A say one with IP address 192.168.2.2 tries to connect to a host on LAN B with IP of 192.168.2.130.
- Packets go to LAN A's router, which "sees" that it is a request for the subnet and does not forward the packet any further up the chain (it could but it would not matter, since on the Y LAN there are no 192.168.2.X hosts).
- The router also sees that it does not know of any host and reports back to the original host that "destination host unknown".

- If the host on LAN A requests an IP address other than those on LAN B, the router will forward the packets up the chain and this continues on to the internet to try and resolve the IP connection.

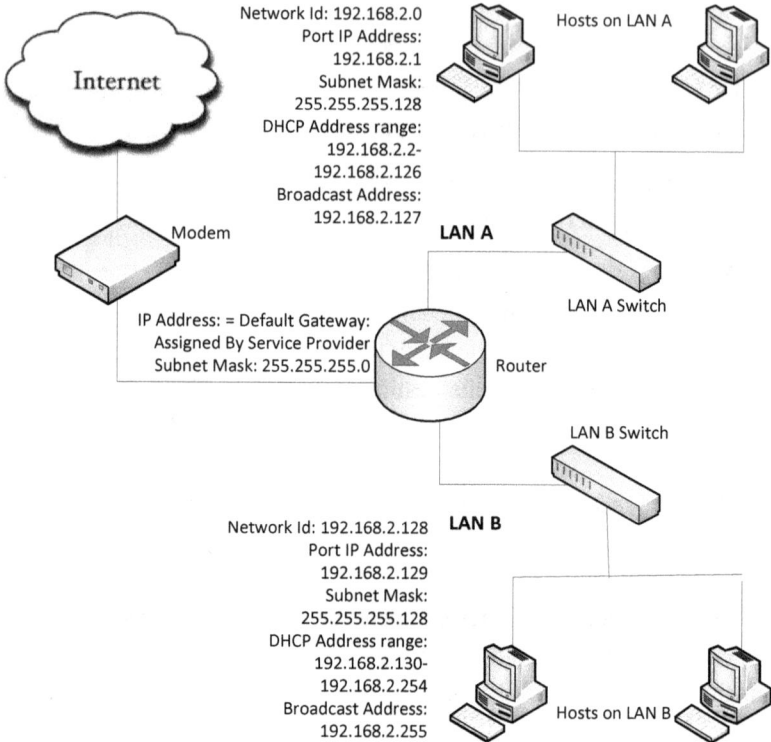

Figure 3.4: Transparent Subnetworks

This leads to two LANs that are physically impossible to talk to each other while both being able to connect to the internet. Such a setup may be desirable in a restaurant where the owner may want to provide free wireless connection for customers, but does not want them to connect to his/her business network. It is also used in industry to isolate machine or production cell networks, adding an extra layer of security. LAN A and LAN B in Figure 3.4 can be implemented on the same router with level 3 LAN ports or layer 2 VLAN capable ports.

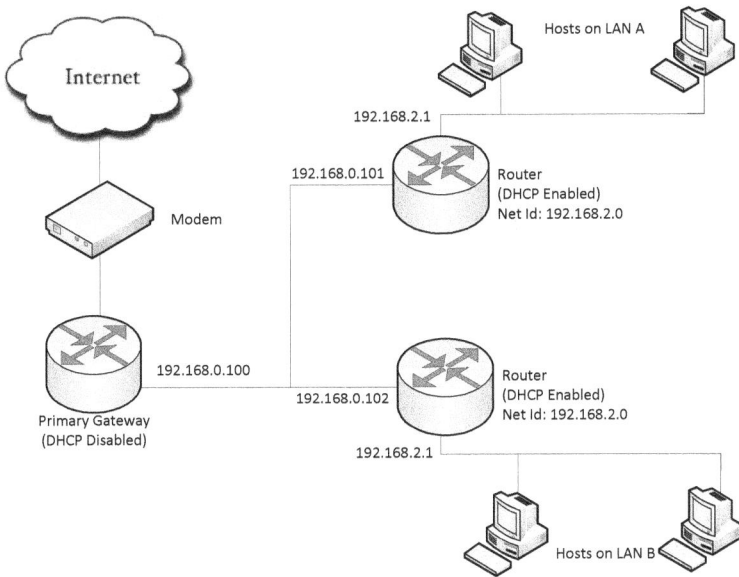

**Figure 3.5: Isolated Subnetworks**

## 3.3    LAN Technologies Specifications

Ethernet is the most widely used LAN technology. But there are other technologies that can be used to implement LANs with widely varied cost and performance implications. For example, Fiber Distributed Data Interconnect (FDDI) is a popular local area networking technology that provides higher bandwidth than Ethernet. Unlike Ethernet and other LAN technologies that use cables to carry electrical signals, FDDI uses glass fibres and transfers data by encoding it in pulses of light. FDDI has ability to detect and correct network problems, such as a break in the network. The network is called self-healing because the hardware can automatically accommodate failure.

LAN technologies are governed by the following standards:

- IEEE 802.3 ( Ethernet)
- IEEE 802.3x ( Extensions)
- IEEE 802.3u ( Fast Ethernet)

The technologies are usually specified as follows: *Number, Word - 2 letters, number,* for example 10Base5-TX 100. This example means the following LAN technology specification:

[49]

- *10* at the beginning means the network operates at 10Mbps.
- *BASE* means the type of signaling (coding) used is baseband.
- *5* indicates the maximum cable length in meters.
- *T* stands for twisted-pair cable.
- *X* stands for full duplex-capable cable.
- *F* stands for fiber optic cable.

Based on this specification, other example of LAN technologies including the following:

- 1000 Mbps, 1000Base-T
- 1000Base-SX 275/550 (depending on the bandwidth and attenuation of the fibre used)
- 1000Base-LX 550/5000 (Fibre)

### 3.3.1 LAN Cable Length

The maximum length of cable segments of a LAN is determined by the cable technology (specification) used. For example:

- Category-5 twisted pair cable (CAT-5) is specified as (100base100-TX), which means that it supports a speed of 100Mb/s to a length of 100 meters.
- Fiber Optic cable specified as 100base2000-FX supports full duplex communication with maximum cable length of 2000 meters (2km), at data speed of 100Mb/s.
- Category-6 twisted pair cable (CAT-6) specified as 1000base100-T supports 100 meters for slower network speeds (up to 1,000 Mbps) and higher network speeds over short distances. For Gigabit Ethernet, 55 meters max, with 33 meters in high crosstalk conditions.
- CAT-6A: Category-6A twisted pair cable (10Gbase-T) supports the same distance CAT-6 cable for 10 Gigabit Ethernet.

Beyond the specified cable length, the LAN may seem to slow down. This is caused by increased packet loss, which is followed by auto retransmit requests. TCP has sophisticated error checking mechanisms that will detect packet loss and triggers re-transmit messages. But if this happens so often, its net effect is reducing the effective throughput (bandwidth). Note that Ethernet based industrial networks do not use TCP for real-time data. This means that their associated error detection methods are not very sophisticated. Therefore packet loss in industrial networks is something that you should void right off the batch by obeying the cable length specifications.

### 3.3.2 Network Infrastructure Devices

The distances specified in Section 3.3 are for single cable spans. Theoretically, a LAN can be extended to cover any distance by joining its spans using repeaters, intermediate switches, and other network infrastructure devices. These devices are described in detail in Section 5.3. Note that the specifications of the devices used in a LAN must much the specification of the cables in order to achieve optimum network performance. For example using cables with full duplex capability with repeaters that have half duplex capability limits the network to half duplex performance.

## 3.4 References

[1] Ishwar Singh, Nafia Al-Mutawaly, and Tom Wanyama, "Teaching Network Technologies that Support Industry 4.0", in Proc. CEEA Canadian Engineering Education Conf., CEEA2015, (Hamilton, Ontario; 31 Many- 03 June 2015), 2015.

[2] Zhihong Lin, Stephanie Pearson, "An inside look at industrial Ethernet communication protocols", *Texas Instruments,* Dallas, Texas, November 2013, Available as of April 12, 2015 from http://www.ti.com/lit/wp/spry254/spry254.pdf

## 3.5    Discussion Questions

**Question 1**

i)      What are the layers of the OSI reference model?

ii)     What is the purpose of each layer?

iii)    What is the encapsulation unit of information at each of the layers?

iv)     What communication device(s) handle the information at each of the layer?

**Question 2**

i)      Draw a network showing the Plant network, Fieldbus, sensor network, sensor network gateway/master, PLCs and Remote I/O

ii)     Draw a network for which all the sensors, PLCs, Operator station and remote IOs have Plant Network (Ethernet) Interfaces.

iii)    Compare the networks in i) and ii).

**Question 3**

Explain why Ethernet based protocols are used at the plant level of the industrial networks architectural hierarchy.

**Question 4: Project Question**

As an automation engineer, IT departments assigned the network 215.161.3.0. You want to subnet the network such that each subnetwork can accommodated 18 to 24 hosts. Determine the following:

i)      Natural subnet mask of the network

ii)     Number of hosts the network accommodated if it is not subnetted

iii)    Subnet mask after subnetting the network

iv)     Number of subnetworks

v)      Maximum number of hosts per subnetworks

vi) Total maximum number of hosts the network can accommodate after subnetting.

vii) List of network addresses and broadcast addresses of the first 4 subnetworks

viii) Draw the implementation of the first four subnetworks of the network in question 3, using a router with layer 3 LAN ports

ix) What is a VLAN?

x) What are the advantages and disadvantages of subnetting?

# Chapter 4 *Ethernet Based Industrial networks*

This chapter focuses on manufacturing automation. Building and electricity substation automation systems are covered in specific chapters. At the plant level of the industrial network architecture, devices share massive amount of information, which requires high capacity networks. Therefore, most plant level networks use Ethernet based protocols such as PROFINET and Ethernet/IP (Figure 2.1), because Ethernet is capable of carrying large amounts of data. In addition, Ethernet based protocols at the plant level make it easy to integrate plant networks and business networks, a feature desired by Industry 4.0. As Ethernet is ubiquitous and cost effective, with common physical links, high speed and determinism, it is poise to becomes the de facto protocol for industrial networks.

It is expected that the data flow between process level sensors/actuators and plant data centers will continuously grow. In line with this growth, this book focuses on adopting Ethernet based protocols as a means of integrating manufacturing, energy, environmental and buildings monitoring, supply chain and customer service centers. Therefore, the book covers Ethernet protocol topics including: Standard TCP/IP Ethernet, Ethernet IP, BACnet, IEC61850, PROFINET, Modbus TCP.

## 4.1    Ethernet IP

Ethernet Industrial Protocol or Ethernet IP was developed by Rockwell Automation. Figure 4.1 compares the Ethernet IP stack, with the TCP/IP stack and the OSI communication reference model. It shows that Ethernet IP has the following characteristics:

- It uses same layers as standard Ethernet.
- Application layer is CIP (Common Industrial Protocol); currently known as the Communication and Information Protocol. The protocol is also used in Devicenet and ControlNet devices.

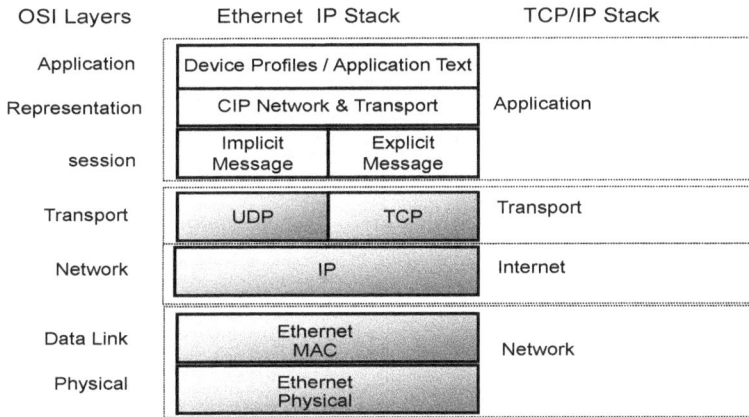

| OSI Layers | Ethernet IP Stack | TCP/IP Stack |
|---|---|---|
| Application | Device Profiles / Application Text | |
| Representation | CIP Network & Transport | Application |
| session | Implicit Message / Explicit Message | |
| Transport | UDP / TCP | Transport |
| Network | IP | Internet |
| Data Link | Ethernet MAC | Network |
| Physical | Ethernet Physical | |

**Figure 4.1: Ethernet IP and the OSI Communication Reference Model**

## 4.1.1 CIP Data Structure

CIP structures device data in classes that are identified by numbers. For example class 01 identifies "Identity Object" which is the information about a device, and class 04 is the device data. The objects have attributes. For example, object "motor drive data" has attributes (identified by a number) such as speed, voltage and current. Each attribute has instances (also identified by a number). For example, motor data – speed attribute has instances such as maximum speed, instantaneous speed, minimum speed, and units. The numbers that specify data classes, attributes, instances and the associated allowed services are referred to as connection points. This information is required in order to correctly configure and program Ethernet IP communication.

## 4.1.2 CIP Messaging

Ethernet IP uses the following forms of messaging defined by CIP standard:

**Unconnected Messaging:** This is used in the process of establishing connections and in infrequent, low-priority messages. Unconnected messages utilize TCP to move messages across Ethernet. Each time the message is sent, the connection is established by asking for connection resources from the Unconnected Message Manager, or UCMM.

**Connected Messaging:** This utilizes resources that are dedicated (reserved) in advance within each node for the messaging purpose. It is used for frequent explicit message transactions or real-time I/O data transfers.

[55]

Moreover, Ethernet IP messaging can be classified as explicit or implicit messaging (Table 4.1 presents a summary of the characteristics of the two communication methods).

**Explicit Messaging:** These messages are sent using TCP/IP. They perform client-server (point-to-point) type transactions between nodes (Figure 4.4). Explicit TCP connected messaging requires setting up a connection between the communicating devices for the entire communication period. This means that resources required to manage the connection must stay reserved for this purpose as long as this connection exists. TCP connected messaging is more reliable than UDP connected messaging, but it requires far more processing power. This is the reason why TCP is usually used for unconnected explicit messaging. In fact explicit TCP unconnected messaging is the compromise between the two extremes of explicit TCP connected messaging and implicit UDP connected messaging. It required less processor power than TCP connected messaging and it is more reliable than implicit UDP messaging. However, its data update rate is far less than the data update rate of implicit UDP messaging. In fact, the data update rate is determined by the devices program logic of the communicating device.

**Implicit Messaging:** Implicit connected messaging utilizes User Datagram Protocol (UDP/IP) for its communication mechanism. This model allows messages to be multicast (this means that the message is targeted to multiple nodes in a network). With Implicit messaging connection, the data fields contain only real-time IO data, and no protocol information. To reduce processing, implicit messaging pre-defines the meaning of the data at the time the connection is established. Although UDP is connectionless and makes no guarantee that data will get from one device to another, its messages are smaller and can be processed more quickly than TCP/IP messages. Moreover, the messages are sent multiple times every second, which minimizes the effects of packet loss. Consequently, Ethernet IP uses UDP/IP to transport I/O messages that typically contain time-critical control data (Figure 3.4). Data reception is scheduled using the requested packet interval (RPI) parameter that specifies the rate at which data is updates.

[56]

**Table 4.1: Two Types of Communication Defined by Ethernet IP**

| CIP Message Type | CIP Communication Relationship | Transport Protocol | Communication Type | Typical Use | Example |
|---|---|---|---|---|---|
| Explicit | Connected or Unconnected | TCP/IP | Request/reply transactions | Non time-critical information data | Read/Write configuration parameters |
| Implicit | Connected | UDP | I/O data transfers | Real-time I/O data | Real-time control data from remote I/O device |

### 4.1.3 CIP Data Types

Ethernet IP supports the following three data transmission types (Note that data types are associated to connections points. Moreover the connection points specify whether the data can be transmitted using explicit or implicit messaging):

- **Information:** Non-time critical TCP/IP data transfers of typically large packet size. The communication sessions are short-lived explicit connections between two devices.
- **I/O Data:** Time-critical UDP/IP data transfers of typically smaller packet size. The communication sessions are long-term implicit connections between one source and multiple devices.
- **Real-Time Interlocking:** Cyclic UDP/IP data synchronization between one producer and multiple consumers.

Table 4.2 presents a summary of Ethernet IP message type as well as connection and data transmission types.

### 4.1.4 Ethernet IP Laboratories

Setting up communication among devices involves the following three knowledge areas: wiring and physical interconnection of devices, configuration of communication parameters, and configuration and programming of communication. These IT knowledge areas are associated with the ISO OSI communication reference model as shown in Figure 4.2.

For Ethernet IP, wiring and physical interconnection of devices involves interconnecting data transmitters and receivers (Data Terminal Equipment – DTE) using hubs, routers, switches, and gateways (Data Communication Equipment -

[57]

DCE). The interconnection process is fairly easy since Ethernet IP uses standard RJ45 connectors. Note that some devices may be connected to the network using wireless technology. For example Figure 4.3 shows that the Ethernet IP DTE used in this lab are interconnected using an Eaton Ethernet switch to form the Ethernet IP Laboratory Unit. These units can be connected into a single LAN by interconnecting the Eaton Ethernet switches through the lab network.

**Table 4.2: Ethernet IP Message and Connection Types**

| ETHERNET/IP Transmission Type | Message Type | Transport Protocol | Description | Example |
|---|---|---|---|---|
| Information | Explicit | TCP or UDP | Non-time-critical Information Data | Read/Write data by message instruction |
| IO Data | Implicit | UDP | Real-time I/O Data | Control real time data from remote I/O device |
| Real-Time Interlocking | Implicit | UDP | Real-time Device Interlocking | Exchange real-time data between two processors |

Once Ethernet IP devices have been wired and interconnected, the next step is to configure their communication parameters. These parameters are associated with the data link and the network layers of the ISO OSI layers (lower layers of the OSI reference model). They include setting the IP addresses, subnet masks and default gateways. Moreover, configuration of Ethernet IP devices involves setting them up to be assigned IP addresses automatically or to have static addresses. This is the focus of LAB A (Appendix A): Configuration of Ethernet IP laboratory. The main objective of this laboratory is to teach students the process of identifying useable IP addresses as well as assigning addresses to the network nodes.

The purpose of laboratory B (Appendix B) is to program communication among Ethernet IP devices from different vendors. The laboratory helps students to learn how to program Ethernet IP implicit UDP messaging, and Ethernet IP explicit TCP connected and unconnected messaging. It focuses on Ethernet IP (upper layers of the OSI reference model). Therefore, it covers knowledge required to carry out network configurations using software tools such as CX Configurator and RsWorks, as well as communication programming tools such as CX Programmer and RsLogix5000, that are specific to Ethernet IP. Ethernet IP laboratory in this book uses the Automation Direct Productivity 3000 PLC and software. Since the communication between the Productivity 3000 PLC and the Eaton ELC-CAENET remote I/O module is based on implicit messaging, memory is allocated automatically to the input, output, and configuration data.

| ISO-OSI Model | TCP-IP/EtherNet IP Model | Knowledge Areas |
|---|---|---|
| Application Layer | CIP | Configuration of Communication, and of Data Access |
| Presentation Layer | | |
| Session Layer | | |
| Transport layer | TCP/UDP | |
| Network layer | IP | Configuration of Communication Parameters |
| Data Link layer | ISO/IEC 8802-3 Ethernet | |
| Physical layer | ISO/IEC 8802-3 | Wiring |

**Figure 4.2: OSI Reference Model, Ethernet IP, and Practical IT Knowledge Areas**

Moreover, there is no need to program the communication, instead data is read and written automatically. Explicit messaging is used for communication between Automation Direct Productivity 3000 PLC and Eaton PowerXL DG1 VFD. Therefore, in this lab students have to configure a client just as they do for implicit messaging, but the reading and writing of data is implemented through explicit messaging instruction in the logic.

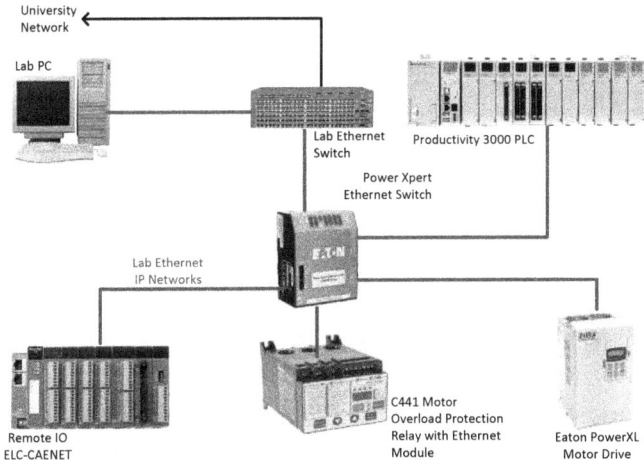

**Figure 4.3: Lab Network**

[59]

## 4.2    PROFINET

PROFINET was developed, and is managed by PROFIBUS International (PI). It is an Ethernet based network solution for the networking of production assets (sensors, actuators, sub-systems and production units) and equipment such as PLCs, DCS and enterprise-wide IT systems. It is fully compatible with and leverages all the features of standard Ethernet. The main differences between standard Ethernet and PROFINET lies in its ability to support real time performance required of industrial automation. In addition, standard Ethernet devices are less able to withstand the harsh industrial environments.

PROFINET is capable of operating in the difficult environments of industry and delivers the speed and precision required by manufacturing plants. It can also provide additional functions, for example Safety, Energy Management and IT Integration. These can be used in combination with the control and monitoring functions.

Other advantages of using PROFINET at the IO level include the following:

- Highly scalable architectures.
- Access to field devices over the network.
- Maintenance and servicing from anywhere (even over the internet).
- Lower costs for production/quality data monitoring.

PROFINET is based on open standard (IEC 61158), and it supports packets of upto1440 bytes at a speed of 1Gbits/s, full duplex communication. Figure 4.4 compares the PROFINET stack with the OSI reference model.

It shows that PROFINET handles the conflicting demands of business networks and plant floor automation using three communications services:

- Standard TCP/IP: This is used for non-deterministic functions such as parametrization, video/audio transmissions and data transfer to higher level IT systems.
- Real Time (PROFINET RT): Here the TCP/IP layers are bypassed in order to obtain deterministic performance for automation applications in the 1-10ms range. This represents a software-based

[60]

solution suitable for typical I/O applications, including motion control and high performance requirements.

- Isochronous Real Time (PROFINET IRT): Here, signal prioritization and scheduled switching deliver high precision synchronization for applications such as motion control. Cycle rates in the sub-millisecond range are possible, with jitter in the sub-microsecond range (increased determinism). This service requires hardware support in the form of ASICs (Real-Time network controllers).

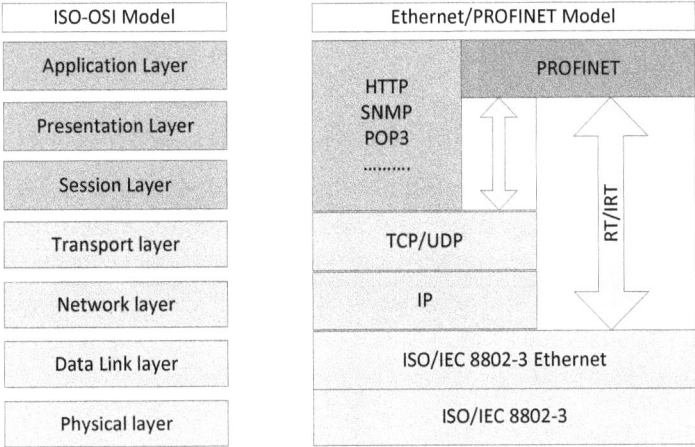

**Figure 4.4: PROFINET and the OSI Reference Model**

All the three services can be used simultaneously. Bandwidth sharing ensures that at least 50% of every I/O cycle remains available for TCP/IP communications, whatever other functionality is being supported. Combined with ruggedized cabling, connectors and Ethernet switches, this means that PROFINET can meet all the needs of automation. Besides, PROFINET can easily be integrated with other fieldbuses by using gateways. Figure 4.5 shows an integration of other networks such as PROFIBUS DP and PA, DeviceNet, AS-I, and HART with PROFINET.

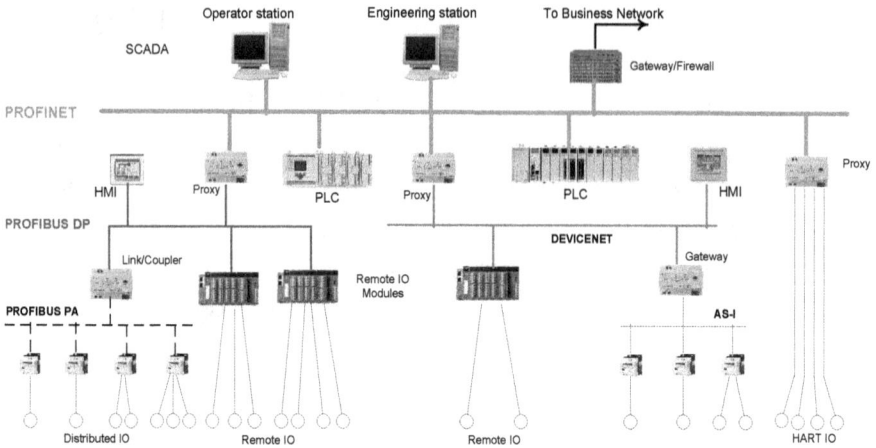

**Figure 4.5: PROFINET - Fieldbus and Plant Level Network**

Figure 4.6 show that the set of skill required to design and physically setup PROFINET networks is the same as the skills require for Ethernet IP networks and indeed any other Ethernet based network. However specific skills are required to program the networks. This special skill set can easily be acquired through work experience. Therefore, labs in only one of the Ethernet based protocols is good enough to teach the principle of such industrial networks.

### 4.2.1    Data Communication

PROFINET treats I/O just like PROFIBUS. An engineering tool associated with the controller is used to obtain information about the I/O devices from a GSD file. After the project is configured in the software, it is downloaded to the controller. The controller can then communicate with the I/O devices. The devices are structured hierarchically as Device, Module, and Channel. Inputs and Outputs are exchanged between the controller and I/O Device as 'cyclic' data. The controller sets the update cycle time. Additional information such as diagnostic data that is not required as frequently is communicated as 'acyclic' data. The communication among PROFINET controllers is based on the principle of producer consumer model with data being updated in cyclic manor. This means the data produced by one device may be consumed by one or more devices.

[62]

| ISO-OSI Model | TCP-IP/PROFINET Model | Knowledge Areas |
|---|---|---|
| Application Layer | HTTP SNMP POP3 ......... / PROFINET | Configuration of Communication, and of Data Access |
| Presentation Layer | | |
| Session Layer | | |
| Transport layer | TCP/UDP | |
| Network layer | IP | Configuration of Communication Parameters |
| Data Link layer | ISO/IEC 8802-3 Ethernet | |
| Physical layer | ISO/IEC 8802-3 | Wiring |

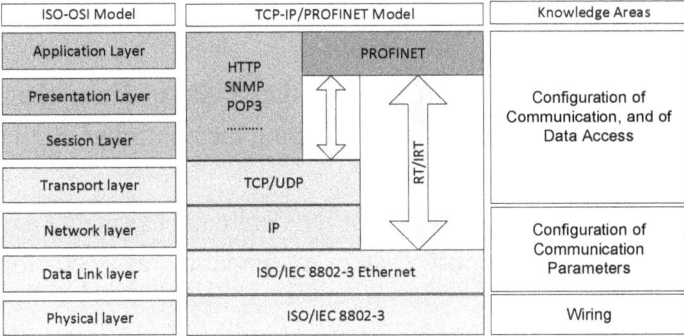

**Figure 4.6: PROFINET and IT Skill Set**

## 4.2.2    Process Automation

Since it is based on Ethernet, PROFINET does not support direct process connectivity. This is because Ethernet was not designed as a hazardous area network, it currently does not have the ability to feed power to field devices like pressure transmitters, and Ethernet based devices require a significant amount of power which may make them unsafe for use in hazardous environments. That is the focus of the IEC 61158-2 standard, and conforming fieldbuses like PROFIBUS PA, and AS-i network remain the only way to connect to field devices in process applications. PROFINET provides a high-speed, high-bandwidth, backbone for PROFIBUS PA (and other fieldbus networks) just like PROFIBUS DP.

PROFIBUS PA and other process automation fieldbuses are connected to PROFINET through associated proxies. The data path via the proxies is fully transparent, thus the instruments on the fieldbus appear directly connected to the PROFINET controller (or DCS) as remote IO. This makes configuration, maintenance and management of an automation system very simple from any part of the enterprise. Another advantage of the proxy approach is that existing field networks do not have to be replaced when upgrading a plant to PROFINET, so investments in skills and inventory are protected as the migration to Industrial Ethernet architectures occurs. Moreover, the same PROFINET backbone can be used to connect other field devices typically found in process applications, for example drives and discrete IO. As most process plants are 'hybrid', that is, they include both process and discrete automation elements, that means only one network, PROFINET, is needed to cover the entire plant.

[63]

Note that PROFINET supports the following three main features needs of process automation:

- *Configuration in Run (CiR)* – or the ability to make changes to an application program without stopping the controller.
- *Time Sync / Time Stamping*
- *Scalable Redundancy* – (PROFINET already has media redundancy, but needs to support system redundancy as well for process applications)

### 4.2.3   Simple device replacement

It is possible to replace IO devices on PROFINET networks without using computers. The controller stores configuration information of devices such that when a device is switched with a similar device, the controller assigns and downloads the configuration information to it, making it operational without the need for external configuration. Since PROFINET uses names, not numbers (e.g. IP address), the replacement device has to have a blank name (reset to factory settings).

### 4.2.4   Wireless

PROFINET supports communication over WLAN (IEEE802.11.1, b, g and n) and over Bluetooth (IEEE802.15.1). In addition PROFINET can be connected to WirelessHART through a standard adapter interface or proxy.

### 4.2.5   Diagnostic

PROFINET and its configuration tools simplify network troubleshooting by providing levels of detail according to user needs, showing device, module, channel and interrupt activity. Furthermore, PROFINET provides the following capabilities:

- Fault events acknowledgement and PROFINET's I&M (Identification and Maintenance)
- Comprehensible naming conventions which means that the user does not have to look up devices by obscure tag names or numbers
- Topology layouts showing the geography of the plant network.
- Integrated web servers are incorporated in automation devices and these mean that a standard Internet browser (such as Explorer or

Firefox) can be used to access diagnostic displays. It also means that engineers don't necessarily have to be on site in order to diagnose a fault as access to plant networks can easily be facilitated over an intranet or even the internet. Automated, event-driven messages can be dispatched by SMS or e-mail to your engineering staff.

- Standard Ethernet tools such as SNMP and Ethereal can be utilized to enhance the available capabilities. There is also easy access to vendor-specific tools for complex devices using the Tool Calling Interface (TCI), which ensures accurate data sharing with PROFINET.

### 4.2.6 Security

The openness of Ethernet guarantees easy access from anywhere in the world using readily available tools such as web browsers. This great advantage has implications for the security of networks across the enterprise.

PROFINET addresses these sometimes critical security issues in various ways, as defined in a special security specification:

- By guarding against errors and improper operation.

- By preventing unauthorized access that could lead to network manipulation or espionage.

- By using proven and certified security standards (e.g. firewalls and VPN).

### 4.2.7 Special Feature of PROFINET

**PROFIenergy:** This intelligent energy management feature switches PROFINET devices into 'sleep' or OFF modes to save on energy usage during:

1. *Expected short breaks* – e.g. lunchtime and shift changes
2. *Planned longer breaks* – e.g. nights and weekends
3. *Unplanned pauses* – e.g. breakdowns, maintenance and upgrades

**Fast Start UP:** The PROFINET's Fast Start Up (FSU) feature makes it possible for an IO device to go instantly into a 'power on' state in response to signals from an IO controller. Such functionality is a high priority for industrial robots with

[65]

Automatic Tool Change (ATC) since it can increase the flexibility of production lines as well as reduce the number of robots per cell. This feature is a result of the following behavior of PROFINET IO devices:

- Use of fixed transmission parameters (only for copper wires), instead of automatic detection, which reduces start up by up to three seconds.
- The network address is not passed to the IO device on every cycle, but only at first start up. Parameters are stored in the IO device memory and re-used. This may save several seconds.
- IO Devices announce their readiness to establish communication instead of waiting for the IO Controller to search. It is possible to save up to one second this way.

**PROFIsafe:** This feature is achieved through the use of additional software layer on top of existing PROFIBUS and PROFINET protocols. It works independently of any automation functions even though it is running on the same network. It can be used with either PROFINET or PROFIBUS, in combination if required, in factory and process automation applications. PROFIsafe supports communication of safety systems that protect equipment, people and the environment. Traditionally, these systems rely on separately wired circuits that are expensive to build, commission, and maintain. PROFIsafe can be used for safety applications up to SIL3 according to IEC 61508 / IEC 62061, or Category 4 according to EN 954-1, or PL "e" according to ISO 13849-1.

### 4.2.8    Design, Installation, and maintenance of PROFINET

*4.2.8.1  Designing*

Design decisions are affected by the following characteristics of PROFINET:

- PROFINET is based on Ethernet. Therefore, it is highly scalable and versatile and can be deployed in Line, Tree, Tree and Branch, Star, and Ring architectures. It supports both IO and peer-to-peer communications.
- It uses Ethernet switches to connect devices. While ordinary switches support PROFINET in difficult environments of most

[66]

manufacturing plants, achieving the extra features that PROFINET offers, means that more sophisticated switches are preferred. In addition, specialized ASIC-based switches are now being fitted into many end devices. Design considerations here include taking into account accumulated switching delays. In linear networks 10 switches in a line is usually the desirable maximum.

- If HMI traffic and data-intensive signals (e.g. vision) are required, it's worth paying attention to overall bandwidth requirements
- PROFINET supports ring topologies whereby, if a cable or device fails, then the system automatically segments itself into a 'line' topology to keep the rest of the system active.

*4.2.8.2 Installation*

Standard 'best practice' in cable installation and maintenance should be followed at all times. Here are a few additional factors that should be borne in mind:

- Remember that the environment you are planning for may be dirty, dusty, electrically noisy and generally unfriendly to data transmissions and infrastructure components of all types.
- Normal twisted pair cabling is suitable for PROFINET. TCP/IP has methods in place to resend telegrams when lost but the timing is not acceptable for industrial use! In other words, electrically noisy environments can easily interrupt your data flows and may cause control malfunctions. In these situations, always use Shielded Twisted Pair cabling.
- Grounding at both ends is best. However, it's not always applicable due to ground loops.
- The need for shielding is independent of the protocol used. All Industrial Ethernets – and indeed fieldbuses – need protection in noisy environments. If you used shielded cable with DeviceNet or PROFIBUS, use shielded cable with PROFINET as well.
- Always use rugged connectors too. Field installable RJ-45 types are available.

[67]

*4.2.8.3 Commissioning*

There are many resources that you can call on to help with commissioning.

- A PROFINET Commissioning Guideline is available, together with a separate Word file, which provides protocols and checklists for individual adaptations.

- Engineering and test tools designed for use with Industrial Ethernet in general, and PROFINET in particular, are available from many sources. Such tools can investigate full Ethernet and PROFINET activity, provide detailed analysis of parameters such as delays and jitter, and time stamp frames for later assessment.

- General-purpose Ethernet tools include the Wireshark software analyzer which uses the Ethernet ports on a PC as the analyzer hardware. Wireshark is license free and acts as a sniffer to analyze the Ethernet traffic

*4.2.8.4 Maintenance*

- Standard Ethernet has a set of diagnostic tools and protocols that will be familiar to office-based technical personnel. These can be utilized in the industrial environment to provide detailed information about lower-level transport-oriented issues such as TCP, UDP and IP activity. They can also support statistical and connection analysis.

- The use of familiar Ethernet protocols means that browser based access to individual devices is possible from any PC in any location, even over the internet. Many PROFINET devices incorporate a web server for this purpose. A browser can also read out information such as device status and configure a device either locally or from a remote site.

- IT protocols familiar from the industrial world include SNMP (for managing components such as switches, and reading statistics and diagnostics). Again these can be used from anywhere in the network. Another familiar protocol is LLDP which is used for mapping network topologies for making device replacement easy.

[68]

- The PROFINET specifications also include a set of specific diagnostic tools which operate at the application layer level. These provide more advanced diagnostics capabilities in standardized formats. Some allow for remote monitoring of networks, and some are intended for on-site use.

- Many devices can be replaced in the field without the need for configuration – no computer is required; you just take the new device out of the box and install it in place of a failed unit.

- OPC servers have network monitoring capabilities using SNMP. They can be used to bring SNMP data to the HMI or to send it automatically to remotely located maintenance support.

## 4.3    Modbus TCP

Modbus TCP is one of the most common Ethernet based industrial networks. It is mainly used for connecting Human Machine Interfaces to controllers and in SCADA systems.

### 4.3.1    Modbus

The Modbus protocol was developed for industrial automation systems in 1979 by Modicon, Incorporated, Now, it is a widely accepted industry standard for transferring discrete and analog I/O information and register data between industrial control and monitoring devices. Developing Modbus devices requires a license, but no royalty payment to its owner. Modbus uses the master-slave (client-server) Media Access Control (MAC) method. Under this method, only one device (the master/client) can initiate transactions (called queries). The other devices (slaves/servers) respond by supplying the requested data to the master, or by taking the action requested in the query. Slaves are usually peripheral devices such as I/O transducers, valves, network drives, or other measuring devices, while masters are usually controllers and PCs. There are devices that may function as both clients (masters) and servers (slaves). Masters can send messages to individual slaves, or can send broadcast messages to all slaves. On the other hand, slaves response to all messages addressed to them individually, but do not respond to broadcast queries.

The Modbus frame structure is the same for requests (master to slave messages) and responses (slave to master messages). Figure 4.7 shows that a master's request consist of a slave address (or broadcast address), a function code defining the requested action, any required data, and an error checking field (the checksum). On the other hand, the slave's response consists of its address, and fields confirming the action taken, any data to be returned, and an error checking field. If a communication error occurs, the slave returns an exception message as its response (see Modbus Exceptions). Traditionally, Modbus messages are transmitted serially and parity checking is also applied to each transmitted character in its data frame. But it is important to note that Modbus is an application layer protocol, it defines the rules for organizing and interpreting data. Generally, it is simply a messaging structure, independent of the underlying physical layer.

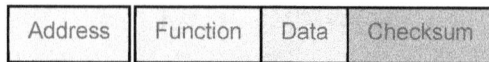

**Figure 4.7: Modbus Message Structure**

### 4.3.2    Modbus TCP/IP

Modbus TCP/IP usually referred to as Modbus-TCP is simply the Modbus RTU protocol with a TCP interface that runs on Ethernet. Figure 4.8 (OSI reference Modal) shows that the Modbus messaging structure is the application protocol that defines the rules for organizing and interpreting the data independent of the data in Modbus TCP. TCP ensures that all packets of data are received correctly, and IP ensure that messages are correctly addressed and routed.

As usual, TCP/IP combination simply provides the transport protocol capabilities, and does not define what the data means or how the data is to be interpreted (this is the job of the application protocol, in this case Modbus). Basically, the Modbus TCP/IP message is a Modbus communication encapsulated in an Ethernet TCP/IP wrapper. Therefore, Modbus inherits many of the strength and weaknesses of TCP/IP, such being multi-master (strength), and not being deterministic (weakness with respect to industrial networks). In practice, Modbus TCP embeds a standard Modbus data frame into a TCP frame, without the Modbus checksum (Figure 4.9).

Figure 4.10 shows that the function code and data fields are absorbed in their original form. Therefore, a Modbus TCP/IP Application Data Unit (ADU) takes the form of a 7 byte header includes the following fields;

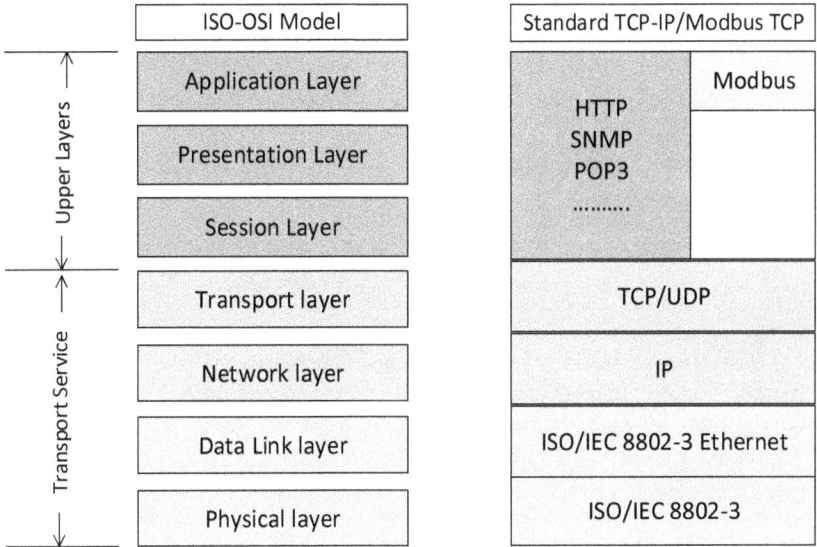

**Figure 4.8: Modbus TCP OSI reference Model**

- Transaction/invocation Identifier (2 Bytes): This identification field is used for transaction pairing when multiple messages are sent along the same TCP connection by a client without waiting for a prior response.

- Protocol Identifier (2 bytes): This field is always 0 for Modbus services and other values are reserved for future extensions.

- Length (2 bytes): This field is a byte count of the remaining fields and includes the unit identifier byte, function code byte, and the data fields.

- Unit Identifier (1 byte): This field is used to identify a remote server located on a non TCP/IP network (for serial bridging). In a typical Modbus TCP/IP server application, the unit ID is set to 00 or FF, ignored by the server, and simply echoed back in the response.

The complete Modbus TCP/IP Application Data Unit is embedded into the data field of a standard TCP frame and sent via TCP to well-known system port 502, which is specifically reserved for Modbus applications. Modbus TCP/IP clients and servers listen and receive Modbus data via port 502. As we can see, the operation of Modbus over Ethernet is transparent to the Modbus register/command structure.

[71]

Application Data Unit ADU

| Address | Function Dode | Data | Checksum |
|---------|---------------|------|----------|

| Function Dode | Data |
|---------------|------|

| 7 Bytes Modbus Application Protocol MBAP Header | | | | Protocol data Unit | |
|---|---|---|---|---|---|
| Transaction Identifier | Protocol Identifier | Length Field | Unit ID | Function Dode | Data |
| (2 bytes) | (2 bytes) | (2 bytes) | (1 bytes) | (1 bytes) | Varies |

Modbus TCP ADU
(This information is embedded into the data portion of the TCP frame)

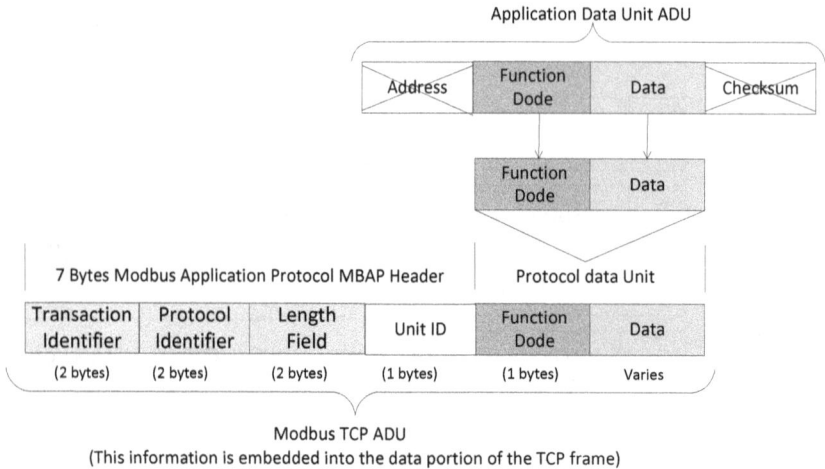

**Figure 4.9: Construction of a Modbus TCP Data Frame**

### 4.3.3 Advantages and Disadvantages of Modbus TCP

Modbus TCP has the following advantages:

- Just as with Ethernet, Modbus is freely available, accessible to anyone, and widely supported by many manufacturers of industrial equipment.

- It is easy to understand and transparent to serial based Modbus networks, making it a good candidate for integrating serial and Ethernet based networks.

- It shares the same physical and data link layers of traditional IEEE 802.3 Ethernet and uses the same TCP/IP suite of protocols, it remains fully compatible with the already installed Ethernet infrastructure of cables, connectors, network interface cards, hubs, and switches.

- It can coexist with other Ethernet based protocol on the same network.

[72]

Main disadvantage of Modbus TCP is its being non-deterministic. Moreover, since it is based on TCP, it cannot be used in hazardous environments that require intrinsically safe networks.

## 4.4    References

[1] Automation Direct, Ethernet IP FAQs – Issue 31, 2015, Available as of April 12, 2015 from http://library.automationdirect.com/ethernetip-protocol-faqs-issue-31-2015/

[2] Open Devicenet Vendor Association, Inc. (ODVA), Ethernet IP Quick Start for Vendors Handbook, Available as of June 30, 2015 from https://www.odva.org/Portals/0/Library/Publications_Numbered/PUB00 213R0_EtherNetIP_Developers_Guide.pdf

[3] Profibus    International    (PI),    PROFINET,    Accessed    from http://us.profinet.com/technology/profinet/

[4] Acromag, Introduction to Modbus TCP/IP, BusWorks 900EN Series - Technical    Reference    –    Modbus    TCP/IP,    Accessed    from: https://scadahacker.com/library/Documents/ICS_Protocols/Acromag%20 -%20Introduction%20to%20Modbus-TCP.pdf

## 4.5 Discussion Questions

**Question** 1

What services does Ethernet IP provide?

**Question 2**

i)    What are connection points?

ii)    Using the following terms: class, service, attribute, and instance, explain the phrase: "CIP data representation is object oriented".

**Question 3**

i)    Explain the terms implicit and explicit messaging with respect to TCP/UDP transport protocol, and connected and unconnected CIP communication relationships

ii)    Using Productivity 300 PLC as an example, compare and contrast the process of configuring explicit and implicit messages

**Question 4**

i)    How would you modify the subnetting in Question 4 in Section 3.6 if the machine network is Ethernet IP?

ii)    Explain your answer for questions 4(i).

**Question 5**

i)    Draw and label the following:

- Layers of the OSI communication reference model, in the box marked "OSI Reference Model" in Figure 4.10.

- Implementation of EtherNet IP with respect to the OSI communication reference model, in the box marked "EtherNet IP" in Figure 4.10.

- Implementation of PROFINET with respect to the OSI communication reference model, in the box marked "PROFINET" in Figure 4.10.

- Implementation of Modbus TCP with respect to the OSI communication reference model, in the box mark "Modbus TCP" in Figure 4.10.

ii) Based on your results in Figure 4.10, what are the two main capability differences among Ethernet based industrial protocols?

| OSI Reference Model | EtherNet IP | PROFINET | Modbus TCP |
| --- | --- | --- | --- |
| | | | |

**Figure 4.10: OSI Communication Reference Models for Ethernet Based Industrial Network Protocols**

# Chapter 5 *BACnet: Building Automation Networks*

There is a growing need to network everyday things to lower costs and environment impact, and to add value and increase comfort. For example, integrating everyday devices such as heaters, air conditioners, refrigerators, lights, pumps, motors, electricity meters, security systems, gates, garage doors, window shades, washers, dryers and elevators enable functionalities as such ones listed below that in turn leads to cost saving, reduced environmental impact and improved quality of life:

- Completion of drying cycle flashes lights inside home.
- Time of day pricing change changes refrigerator defrost cycle.
- Motion alarm triggers lamp and generates remote alarm message.
- Away mode of thermostat changes hot water temperature to save energy.
- Activation of TV dims room lights.

## 5.1 BACnet

Networking building automation devices is not a new idea, but the focus has always been to integrate so as to deliver a certain building functionality cheaply by reducing the amount of wiring. An example of such systems is the Heating Ventilation and Air Conditioning (HVAC). Moreover, integration of building automation systems have traditionally been done using proprietary networks which cannot be depended upon to interconnect everything in buildings on one hand, and to connect the buildings to the insfrustucture system on the other. Now the focus of networks is to connect buildings to infrastructure so as to achieve the benefits of IOT across industries. Since the general insfrustructure systems use Ethernet based protocols, the building automation community is increasing adapting Ethernet based protocols such as BACnet to provide the backbone for the communication of their building automation systems. Therefore from here on this book focuses on the descrption of the BACnet protocol and how it can be used to integrate buildings, manufacturing, energy, environment, and business systems to form a true and complete Internet of Things.

BACnet stands for Building Automation and Control Network. The protocol was introduced in the mid-1990s by a committee of Canadian and American engineers and other building automation specialists. The committee was spearheaded by the American Society of Heating, Refrigerating, and Air-Conditioning Engineers (ASHRAE). Many high speed and deterministic industrial automation protocols such as AS-i that were developed around the same time with BACnet were based on the three layer reduced Open Systems Interconnection (OSI) model. But BACnet includes the network layer, making it a four layer protocol as shown in Figure 5.1. This was done to ensure data routing among the following five data links supported by BACnet: Ethernet, ARCNET, Master-Slave/Token-Passing (MS/TP), Point-to-Point protocol (PTP), and LonTalk.

Figure 5.1 shows that at the lower levels of the OSI model, namely: the physical layer which is concerned with sending symbols representing binary data across a medium; and the data link layer which defines how stations are addressed and how data in form of frames are sent between stations within a Local-Area-Network (LAN); BACnet shares standards with other industrial networks such as Ethernet IP, Profibus, and LonWorks. But at the network layer, the protocol defines its specific standards that allow routing data among five data link which otherwise have stations with incompatible data links and cannot communicate to one another at that level. Moreover, it defines application layer objects and properties which specify the meaning of the data that is sent among stations.

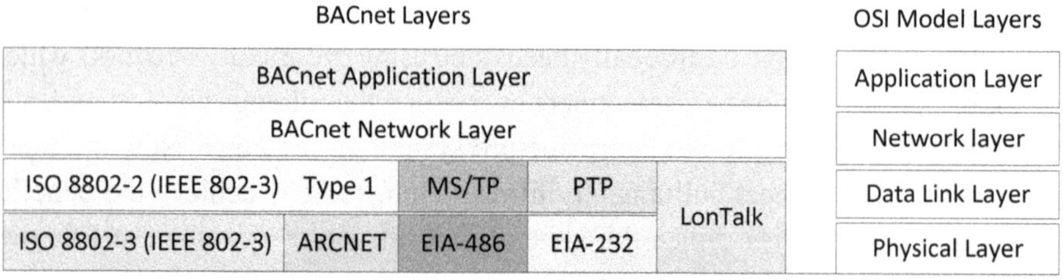

| BACnet Layers | | | | | OSI Model Layers |
|---|---|---|---|---|---|
| BACnet Application Layer | | | | | Application Layer |
| BACnet Network Layer | | | | | Network layer |
| ISO 8802-2 (IEEE 802-3) | Type 1 | MS/TP | PTP | LonTalk | Data Link Layer |
| ISO 8802-3 (IEEE 802-3) | ARCNET | EIA-486 | EIA-232 | | Physical Layer |

**Figure 5.1: BACnet and the OSI Model**

BACnet Application Layer made up of two closely related parts, the first part being a model of the information contained in building automation devices, and the second part being a group of functions or "services" used to exchange information among building automation devices. BACnet does not define the internal design and configuration of devices, but it defines a set of abstract data structures called

[78]

"objects", the properties of which represent the various aspects of the hardware, software, and operation of the device. BACnet objects provide a means of identifying and accessing information without requiring knowledge of the details of a device's internal design. The communication software in devices interpret requests for information about these abstract objects and translate those requests to obtain the information from the real data structures inside the device. Collectively, these objects provide a "network visible" representation of the BACnet device. BACnet defines 18 standard object-types presented in Table 5.1. Moreover, Table 5.2 presents all BACnet services as well as their classification [5].

**Table 5.1: BACnet Standard Object Types**

| Analog Input | Event Enrollment |
|---|---|
| Analog Output | File |
| Analog Value | Group |
| Binary Input | Loop |
| Binary Output | Multi-state Input |
| Binary Value | Multi-state Output |
| Calendar | Notification Class |
| Command | Program |
| Device | Schedule |

Most names of BACnet objects identify the purpose of the objects. But a few that require some explanation are described as follows [5]:

**Calendar** represents a list of dates that have special meaning when scheduling the operation of mechanical equipment, e.g a list of weekend days or holidays.

**Command** represents a multi-action command procedure. For example a sequence of shutting down a group of devices.

**Device** is a logical component that contains general information about a physical device. Such information includes vendor name, model name, location, protocol version supported, and object-types supported.

**Event Enrollment** provides one way to define alarms or other types of events and to indicate who should be notified when they occur. Some objects (Analog Input, Analog Output, Analog Value, Binary Input, Binary Output, Binary Value, and Loop) contain optional properties to support intrinsic event reporting capability and do not need to use Event Enrollment objects.

[79]

**Group** provides a shorthand way to read several values in one request. For example, it might be used to simultaneously update several fields on an operator graphic display.

**Loop** can be used to represent any feedback control loop, which is some combination of proportional, integral, or derivative control.

**Notification Class** provides a way to manage the distribution of alarm or event notifications that are to be sent to multiple devices.

**Table 5.2: BACnet Application Layer Services**

| Alarm and Event Services | Object Access Services | Remote Device Management Services |
|---|---|---|
| *AcknowledgeAlarm* | *AddListElement* | *DeviceCommunicationControl* |
| *ConfirmedCOVNotification* | *RemoveListElement* | *ConfirmedPrivateTransfer* |
| *ConfirmedEventNotification* | *CreateObject* | *UnconfirmedPrivateTransfer* |
| *GetAlarmSummary* | *DeleteObject* | *ReinitializeDevice* |
| *GetEnrollmentSummary* | *ReadProperty* | *ConfirmedTextMessage* |
| *SubscribeCOV* | *ReadPropertyConditional* | *UnconfirmedTextMessage* |
| *UnconfirmedCOVNotification* | *ReadPropertyMultiple* | *TimeSynchronization* |
| *UnconfirmedEventNotification* | *WriteProperty* | *Who-Has* |
| | *WritePropertyMultiple* | *I-Have* |
| | | *Who-Is* |
| **File Access Services** | **Virtual Terminal Services** | *I-Am* |
| *AtomicReadFile* | *VT-Open* | **Security Services** |
| *AtomicWriteFile* | *VT-Close* | *Authenticate* |
| | *VT-Data* | *RequestKey* |

## 5.2    BACnet Data Links

This section describes the six data links supported by BACnet.

### 5.2.1    BACnet/Ethernet

BACnet/Ethernet is based on IEEE 802.3 standard. The standard refers to part 3 of the Institute of Electrical and Electronics Engineers' 802 standard entitled Carrier Sense Multiple Access with Collision Detection (CSMA/CD) MAC method and physical layer specifications.

Ethernet supports a data link that can operate at speeds of 10 Mbps, 100 Mbps and 1 Gbps and its physical layer includes several copper and fiber options such as

[80]

10BASE2, 10BASE5, 10BASE-FL, 100BASE-TX, 100BASE-FX, and 1000BASE-T. In fact of all the BACnet data links, Ethernet provides the greatest speed. Any of the Ethernet physical media can be combined in one piece of equipment and each is compliant with BACnet/Ethernet. Note that BACnet/Ethernet is not the same as BACnet/IP Ethernet. BACnet/Ethernet accomplishes LAN addressing using the Ethernet's Media Access Control (MAC) address. The MAC address is the 48-bit worldwide unique value given to every Ethernet controller chip and not the 32-bit IPv4 address we would expect with an IP/Ethernet device. Bacnet/Ethernet achieves data routing through the BACnet Network layer standard.

### 5.2.2    Bacnet/IP

Due to increasing popularity of Ethernet based industrial protocols as well as the need to integrate building with other infrastructure such as water and electricity utilities, the BACnet community was compelled to modify their protocol to work in the IP world. This is because most Ethernet based industrial protocols used TCP/IP among other standards.    Annex J of the BACnet standard describes BACnet/IP (B/IP), a standard that enables BACnet packets to be routed on the standard TCP/IP network. BACnet/Ethernet and BACnet/IP use MAC addresses for data link, but for the BACnet/IP, IP addresses are needed. BACnet/IP uses the same four-layer OSI model as shown in Figure 1. But instead of using BACnet Network layer, it uses IP. To support high speed data communication, the BACnet community registered a range of 16 UDP port numbers as hexadecimal BAC0 to BACF. TCP is not used because most BACnet packets are not fragmented. In a few cases where packets are fragmented, they are reassembled by the application. Note that BACnet/IP devices can directly attach to the building's IP infrastructure in most cases using Ethernet [1, 3].

### 5.2.3    Bacnet ARCNET

ARCNET was developed by Datapoint Corporation and is a token passing, low cost alternative to Ethernet. It operates at the respectable speed of 2.5 Mbps. It is claimed to be the oldest commercially available LAN, with nearly 3 million nodes installed. The protocol is deterministic, meaning that it is possible to place a bound on the maximum time that a device could have to wait before having a chance to transmit a message. This is a very important feature in building automation applications [2]. At the physical layer, ARCNET uses twisted pair or coaxial cable.

[81]

### 5.2.4    BACnet MS/TP

BACnet Master-Slave/Token-Passing (MS/TP) is a token passing data link with baud rates of 9.6 kbaud to 76.8 kbaud. This flavor of BACnet remains popular because it uses very common physical layer called 2-wire EIA-485, and EIA-485 transceivers are relatively inexpensive and are typically found in low-cost controllers. In fact lavatory controllers and programmable thermostats are usually BACnet MS/TP. These networks are normally connected to the IP infrastructure using BACnet MS/TP to BACnet/IP routers which are then connected to Ethernet switches. The switches are not protocol-aware and only facilitate connections to the IP infrastructure [5].

### 5.2.5    BACnet LonTalk

LonTalk is a proprietary protocol developed and owned by Echelon Corporation. It is a seven-layer protocol implemented on a single chip called a neuron. No physical medium is specified in the LonTalk protocol. Instead, a transceiver interface is defined. A wide variety of physical media can be used by selecting appropriate compatible transceivers. The LonTalk protocol is low-cost, low-speed, and has a very limited (128 octets) message size. Proprietary development tools from Echelon are required to implement the protocol. Devices based on the LonTalk protocol are, in general, not interoperable except when external constraints are applied. In BACnet the LonTalk protocol is used solely as a means to transfer data from one device to another just like all the other BACnet LAN options. Moreover, interoperability is achieved in BACnet LonTalk devices by using the same application and network layer protocols as all other BACnet devices [2, 4].

### 5.2.6    BACnet PTP

BACnet Point to Point protocol provides for inter networked communications over modems and voice grade phone lines. Moreover, supports direct cable connections using the EIA-232 (RS 232) signaling standard. However, its speed is limited the range of 9.6 kbit/s to 56.0 kbit/s.

## 5.3    Equipment that Support BACnet Internet

BACnet is designed to setup a network of building networks. Such a network is generally referred to in literature as an Internet. Developers of BACnet believed that buildings provide a variety services that poise various networking requirements which are difficult to satisfy by a single communication protocol. But they also realized the need to integrate these network protocols through common network and application layers (Figure 4.1). In order to achieve different levels of network integration a multicity of Data Communication Equipment (DTE) are used in BACnet networks. Figure 5.2 shows the four-layer communications reference model that BACnet uses as well as the DTE associated with each layer [1, 3]:

| Layer | Data Encapsulation | Communication Device |
|---|---|---|
| Application | Applications Data | Gateway |
| Network | Packets | Router |
| Data Link | Frames | Bridge/Switch |
| Physical | Physical | Hub |

**Figure 5.2: BACnet Layers, Data Encapsulation and Associated DTEs**

**Repeater:** The physical layer is concerned with the transmission and reception of symbols representing binary data sent across the medium. A repeater is used to filter, amplify and resend an incoming signal. This allows the signals to be received at distances it would otherwise not reach. An example of a repeater is an Ethernet repeating hub. Repeaters are simple devices that do not understand protocols or data link under which they operate.

**Bridge:** A bridge is used to connect two or more network segments of the same data link together. Ethernet switch is a good example of a bridge. A bridge receives the entire frame before forwarding it. It also has the ability to learn the location of hosts leading to a more efficient frame forwarding process. An Ethernet switching hub is fundamentally different from an Ethernet repeating hub. Unlike a repeating hub, an Ethernet switch has the ability to terminate an individual collision domain at each of its port, resulting in achieving much greater distance. Therefore, cascading switches does not limit the physical Ethernet network like repeating hubs. Furthermore a switch can operate in full-duplex mode — assuming its link partner is full-duplex compatible. Full-duplex may effectively double the throughput.

**Router:** The router operates at the network layer of the OSI reference model. It is used to route packets between networks that use the same network layer. Therefore

[83]

it is the most important DTE in BACnet because it is the one that interconnects the various flavors of the protocol. The most common example of a router is the IP router. But note that a BACnet router is not necessarily an IP router. A BACnet router understands the BACnet Network Layer protocol and not the Internet Protocol unless it supports BACnet/IP. A BACnet router attaches two or more BACnet data links together to form one BACnet internetwork. Unlike with an IP router, where Ethernet is typically present on each side of the router, a BACnet usually has different physical media on each side. This is because of the different data links BACnet supports.

**Gateway:** Loosely applied, the term gateway and router are usually used interchangeably. But in their strict sense a gateway operates at the application layer while a router operates at the network layer. A gateway works on messages sent between two different (incompatible) application layers. Therefore a gateway uses custom software to interpret data representation of one application layer for another application layer so that meaningful data can be exchanged between the two devices. An example of a gateway is the device that enables data exchange between the BACnet application layer and the Modbus application layer. But converting Modbus serial to Modbus TCP is not done at the data link layer since the application layers are compatible and Modbus serial does not have the network and transport layers. However, the term gateway is usually loosely applied to this situation as well.

## 5.4    Internet

Like in many other industries, remote access and cloud computing is gaining popularity in building automation. This is due to the increasing need to make building as well as their services part of the Internet of Things, a concept that is believed to increase efficiencies, lower cost, reduce environmental impact and increase quality of life. But in order for devices on BACnet LANs to gain remote access to or from the Internet, an IP Router is needed. For security, the IP router also functions as a state-full firewall. On the LAN side of the router is the BAS which is treated as an intranet and the WAN side is the Internet [3].

## 5.5    References

[1] ASHRAE, Standard 135-1995: BACnetTM - A Data Communication Protocol for Building Automation and Control Networks. American

Society of Heating, Refrigerating, and Air-Conditioning Engineers. Atlanta, Georgia, USA. (1995).

[2] Bill Swan "Internetworking with BACnet: A first look at networking in BACnet", Alerton Technologies Inc., Available as of April 12, 2015 at http://www.bacnet.org/Bibliography/ES-1-97/ES-1-97.htm

[3] Control Solutions Inc., "Connecting BACnet Devices to an IP Infrastructure", Building Automation.com, Minnesota, 2009, Available as of April 12, 2015 at http://www.automatedbuildings.com/news/mar09/articles/cctrls/0902200 20828cctrls.htm

[4] LonTalk® Protocol Specification Version 3.0. Echelon Corporation, 4015 Miranda Ave., Palo Alto, CA 94304.

[5] Steven T. Bushby, "BACnet - A standard communication infrastructure for intelligent buildings", National Institute of Standards and Technology, USA, Automation in Construction, Vol. 6 No. 5-6, 1997, p. 529-540

## 5.6    Discussion Questions

**Question 1**

What are the advantages and disadvantages of BACnet?

**Question 2**

Describe the five data links of BACnet.

**Question 3**

What is the difference between BACnet/Ethernet and BACnet/IP?

**Question 4**

Explain how you would connect BACnet/Ethernet to the Internet for remote data access and system control.

**Question 5**

Using the OSI reference model, compare and contrast Ethernet IP and BACnet.

**Question 6**

Using the OSI reference model, explain the differences between a router and a gateway.

**Question 7**

Compare and contrast an Ethernet and a BACnet router.

**Question 8**

Describe how a Modbus serial network is interconnected with a BACnet/Ethernet network.

**Question 9**

Describe how a Modbus TCP network is interconnected with a BACnet/Ethernet network.

# Chapter 6 *IEC 61850: Electrical Substation Automation Networks*

The IEC61850 is the new international standard for communication within electrical substation aimed at promoting interoperability between devices and systems from different vendors. The standard achieves this through providing the following services [1]:

**Abstract object modelling:** The standard provides an object model defining data which can be provided or requested by functions that reside in equipment (physical devices). The standard does not define these functions, but it requires that devices from different manufacturers be able to exchange the associated data. This means that the services provided by the IEC61850 standard are independent of the implementation technology. Note that configuration files are required to enable devices from different manufacturers to communicate.

**Communication services:** The standard provides the specification for data transfer among Intelligent Electronic Devices (IED) within the substation.

While there are many services provided by the IEC61850 standard. This book focuses on its services, object modelling, and communication.

## 6.1    IEC61850 Data Object Modelling

The IEC61850 data object model starts with the Physical Device (PD), which is the device that connects to the network (Figure 6.1). Physical devices contain one or more logical devices which represent groups of device functions such as protection, control, metering, or human interface functions. In each Logical Device (LD) there is at least one Logical Node (LNs), and the LN breaks down further into data attributes that define the name, format, range, and representation of possible values related to some power system function such as status or measurement.

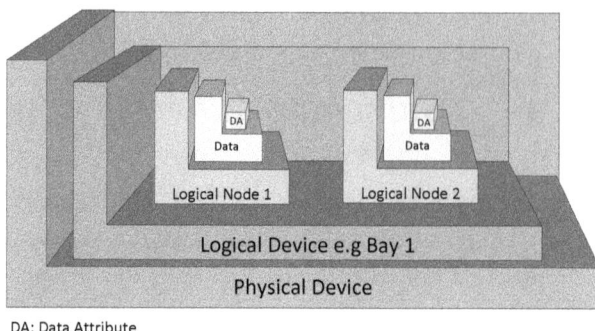

DA: Data Attribute

**Figure 6.1: Data Object Model of IEDs in the IEC61850**

IEC61850 specifies 13 groups of 91 LNs, and Table 6.1 presents some example of the LN groups. Table 6.2 shows that the first letter of the LN name identifies the group. Moreover, the table shows the naming of some of the LNs (objects) related to the protection function [4].

**Table 6.1: Some Example Logical Node Groups**

| Logical Group | Name |
|---|---|
| P | Protection |
| C | Control |
| M | Measuring |
| X | Switchgear |

**Table 6.2: Naming of Some Logical Nodes of the Protection Group**

| LNname | Protection Function |
|---|---|
| PDIF | Differential |
| PDIR | Direction comparison |
| PIOC | Instantaneous overcurrent |
| PTOC | Time overcurrent |

LN is a very important element of the IEC61850 data model because typical functions of the standard are executed by the exchange of information among different LNs. For example, Figure 6.2 shows the participating LNs to execute a distance protection function, namely: distance protection LN (PDIS), current transformer LN (TCTR), voltage transformer LN (TVTR) and the circuit breaker LN (XCBR) [4].

[88]

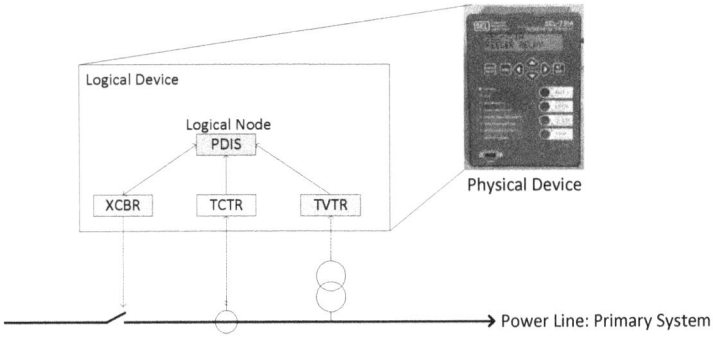

**Figure 6.2: Example of Logical Nodes Participating in a Protection Function**

## 6.2 IEC61850 and the OSI Communication Reference Model

The OSI model of the IEC61850 standard in Figure 6.3 shows that besides defining the data and data structure of IEDs, the standard also defines a hybrid of communication protocols over which the defined data is shared among IEDs. The use of multiple communication protocols emanates from the appreciation that IEDs have multiple data communication requirements that cannot be satisfied by a single protocol.

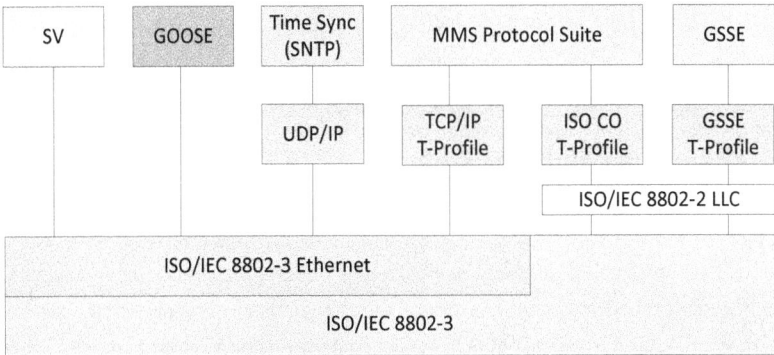

**Figure 6.3: IEC61850 OSI Reference Model**

Features of the IEC61850 OSI communication reference model:

**Sample Values (SV):** A method for transmitting sampled measurements from transducers such as CTs, VTs, and digital IO. It enables sharing of I/O signals among IEDs, and it supports 2 transmission methods: Multicast service over Ethernet and Unicast (point-to-point) service over serial links.

[89]

**Generic Substation Status:** A fast and reliable system-wide distribution of input and output data values. It is based on a publisher/subscriber mechanism and it supports simultaneous delivery of the same generic substation event information to more than one physical device through the use of multicast/broadcast services. IEC61850 defines two types of Generic Substation Status communication protocols, namely GOOSE and GSSE.

**Generic Substation Status Event (GSSE):** It provides an express way of substation level status exchange. It is similar to GOOSE, but it only supports a fixed structure of the data to be published, and not datasets that are created by the programmer like in GOOSE messaging. It is based on multicast and an extension of event transfer mechanism in UCA2.0. GSSE messages are transmitted directly over IEC/ISO 8802-2 and 8802-3 using a similar mechanism to GOOSE messages (IEC 61850-7-1 Clause 12.2, IEC 61850-8-1 Clause 6.4). As the GSSE format is simpler than GOOSE it is handled faster in some devices. GSSE is being progressively superseded by the use of GOOSE and support for it may eventually disappear [5].

**Generic Object Oriented Substation Event (GOOSE):** Used for fast transmission of substation events, such as commands, alarms, and indications, as messages. A single GOOSE message sent by an IED can be received by several receivers. It takes advantage of Ethernet and supports real-time behavior. For examples: tripping of switchgear and providing position status of interlocking.

## 6.3    Advantages and Disadvantages of IEC61850

IEC61850 brings the following strength to electricity substation automation:

- Improved performance.
- Flexible: ability to interconnect electrical data network with the business network and with other industries that may need to receive real-time data from the electricity industry.
- Easier to add more points, devices or IEDs.
- Easier to maintain using modern tools based on TCP/IP technologies.

However the main disadvantages of IEC61850 is the high cost of compliant devices.

## 6.4    IEC61850 Devices and Communication Architecture

IEC61850 defines communication at two levels, namely the station level and the process level. This leads to a network architecture that has two buses: the station bus (IEC61850-8-1) and the process bus (IEC 61850-9-2) (Figure 6.4). According to IEC61850 the use of the process bus is not mandatory. But using only one communication bus – the station bus - represents partial implementation of IEC61850.   In this architecture, the interface with the primary equipment is identical to that of conventional substations. There are hardwired connections between the following devices:

- Instrument transformers and the analog inputs of the IEDs.
- Auxiliary contacts of the breakers and the IED opto-inputs.
- IED digital outputs and the process control inputs.

Communication over the station bus local area network is used for message exchange between IEDs and Supervisory Control And Data Acquisition (SCADA) systems. Laboratories 2A and 2B focus on the exchange of the Generic Object Orientated Substation Events (GOOSE) messages among IEDs. Other messaging is available for the generation and transfer of reports over Manufacturing Messaging Specification (MMS). This is the partial focus of laboratories 3A and 3B.

Full implementation of IEC61850 uses both the station and process buses. The interfaces among all devices in the system in this case are based on bus communications, with the use of copper cables limited to DC or AC power. The process bus is for transferring Sampled Values (SV) between intelligent primary equipment and IEDs. SV are time critical data such as sampled values of current and voltage signals. The full implementation of IEC61850 requires the use of non-conventional instrument transformers and other intelligent sensors as well as communication-interfaced switchgears. These technologies are not yet well developed, thus most substation automation systems available today utilize only partial implementation of the IEC61850.

The SEL-751A relay is designed for the partial implementation of the IEC61850 standard. It has the ability to provide the following services to substation monitoring and control networks within the IEC61850 standard:

1. Collection and digitization of data from CTs and VTs.
2. Collection and digitization of status data.

[91]

3. Operation of breakers.
4. Metering to display the present values of current, voltage, operation status, RTD measurements, and feeder load. The relay also meters real, reactive, and apparent power supplied to feeder loads.
5. Supply of data to other components of the network (e.g. other IEDs, HMIs and Control Center) through dual (redundant) 10/100 BASE-T Ethernet.

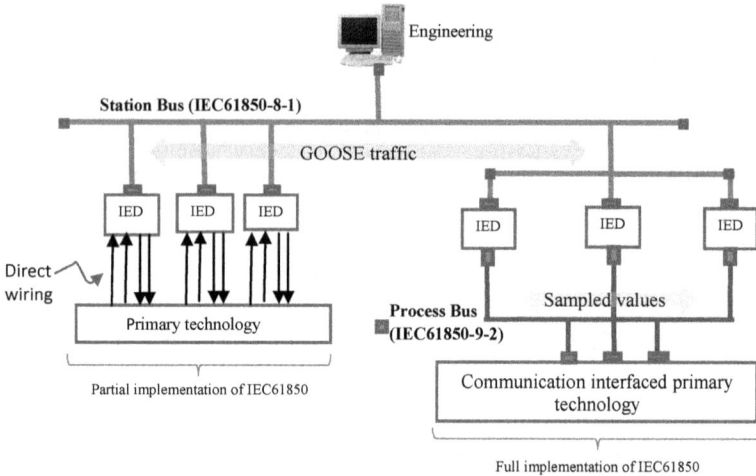

**Figure 6.4: Substation Communication Architecture**

## 6.5 IEC61850 Devices and GOOSE Messages

GOOSE is an IEC61850 object for high-speed control messaging. It is used to replace the conventional hardwired logic necessary for intra-IED coordination with high-speed station bus communications. GOOSE messaging supports the exchange of a wide range of common data organized in datasets.

Upon detecting an event, the IEDs use multi-cast transmission to broadcast GOOSE messages containing data in the associated dataset. This data may be configuration, status, controls, or measured values. Only the devices that have subscribed to specific data in a GOOSE message react to it. The reaction of each receiver depends on its configuration, logic programming and functionality, such as tripping of switchgear, starting of disturbance recorder, or providing position indication for interlocking. GOOSE messages are re-transmitted multiple times. The delay between transmissions increases with each re-transmission.

[92]

## 6.6    IED Configuration Tools

The tools that are used to engineer substation automation systems are not governed by the IEC61850 standard. Therefore, each manufacturer uses proprietary software tools to support the engineering of the substation automation functions. These configuration tools translate the IED capabilities (e.g. functionality, data communication, events and alarms) to the Substation Configuration Description Language (SCL), which is defined in the standard. The SCL allows the exchange of information among different manufacturers' configuration tools. It also enables transfer to the configuration from the tools to the IEDs. There are four types of SCL files, namely:

- *IED Capability Description (.ICD) file*: The ICD file describes the capabilities of an IED, including information on LN and GOOSE support.
- *Configured IED Description file (.CID)*: The CID file, of which there may be several, describes a single instantiated IED within the project, and includes address information.
- *System Specification Description (.SSD) file*: The SSD file describes the single-line diagram of the substation and the required LNs.
- *Substation Configuration Description file (.SCD)*: The SCD file contains information on all IEDs, communications configuration data, and a substation description. Each proprietary configuration tool must be able to import this SCD file and extract the information needed for the IED.

Our IEC61850 laboratories are based on the Schweitzer Engineering Laboratories (SEL) IEDs Therefore, we describe the following two tools used to configure SEL IEDs, and engineer substation automation functions based on SEL devices: *AcSELerator Quickset* and *AcSELerator Architect*.

### 6.6.1    AcSELerator Quickset

AcSELerator Quickset provides GUI for setting up the communication parameters of SEL devices. It also enables the following:

- Design logic control equations. Some of these equations determine the response of IEDs when GOOSE messages are received.
- Verification of logic control equations and event reports with an integrated logic simulator.
- Management, creation, cropping, and reading device settings database manager.

[93]

- Selecting the settings that need to be shown to users.
- Design boolean and mathematic control expressions.
- Assignment of easily recognizable alias names to settings.
- Access to HMI for quick monitoring of electrical system parameters.

### 6.6.2 AcSELerator Architect

AcSELerator Architect provides a GUI (Figure 6.5) for the selection, editing, and creating of IEC 61850 GOOSE messages for substation protection, coordination, and control schemes. The GUI also supports the configuration of reports that are used in SCADA system through MMS communication.

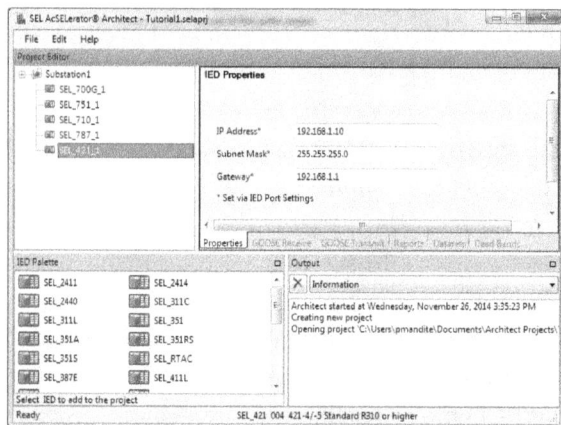

**Figure 6.5: SEL AcSELerator Architect Project Editor**

The following steps are carried out to configure GOOSE messages, through the GUI shown in Figure 6.5:

- Select IEDs for the substation project from the IED palette. The palette is a collection of the default ICD descriptions of the available IEDs.
- Configure the communication parameters of all participating IEDs.
- Configure the data sets of each IED.
- Set the data sets for the transmit GOOSE messages of IEDs.
- Configure the GOOSE messages IEDs can receive. Opening the GOOSE receive tab on the Project Editor reveals the available receive data from each IED in the project (or substation), structured as partly illustrated in Figure 6.6. For example, to receive the status (stVal) of circuit breaker

BS1XCBR2, the attribute *stVal* is mapped to some logic variable defined in the IED. This variable is then read and used by functions in the device.

- Response to the received GOOSE messages is determined by the logic in each IED. This logic is configured using AcSELerator Quickset software.
- Generate specific IED Capability Description (CID) file for each IED and transfer it to the associated IED via File Transfer Protocol (FTP) to IEDs. Note the IEC 61850 has to be is enabled in the IED through the AcSELerator Quickset software.

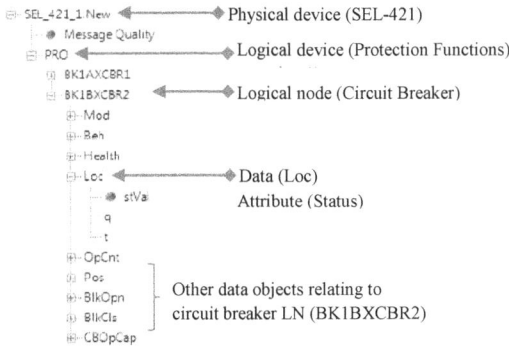

**Figure 6.6: SEL AcSELerator Architect Data Structure**

## 6.7 IEC61850 Laboratory

There are four laboratories on IEC61850 in this book. Laboratory 2A (Appendix C) focuses on the configuration of IED, laboratory 2B (Appendix D) deals with GOOSE messaging, and laboratories 3A and 3B cover using MMS protocol to access IED data. The laboratories demonstrate the need to use GOOSE to send small real-time data, and to use MMS to access large amounts of data required by SCADA systems. The Schweitzer Engineering Laboratories (SEL) SEL 751A relay is used in all IEC61850 laboratories in this book.

Schweitzer Engineering Laboratories (SEL) produces many devices for power systems management. The SEL-751A feeder protection relay is used to protect distribution power lines. Each branch of an electrical distribution system should be protected by fuses and provided a means of disconnect, in accordance with electrical safety codes. In residential systems, both requirements are achieved through the use of breakers. In commercial and industrial systems the means of disconnect should be within a distance of a device being protected. This ensures

[95]

that electricians have visual contact with the disconnect system while working on the device, so that others, or control systems, cannot make the device live (powered) without their prior knowledge and consent. The use of electronic relays such as the SEL751A achieves all these requirements.

A typical distribution system is portrayed in Figure 6.8 (at end of Chapter), which includes diesel generators for emergency backup power. While the arrows of the 34,500 Volt feeders are backwards, it can be seen that two separate lines of utility feeders, as well as two separate diesel generator feeders can be used to provide power to the un-indicated facility at the lower portion of the figure. Only one feeder line can be connected at any given time in the outdoor substation, to provide one source of power to the step-down substation.

Greater power efficiency is achieved at higher voltages, so power systems that distribute power at higher voltages ultimately require the voltages to be stepped down to usable levels prior to entering facilities. The SEL feeder protection relay is used to protect transformers and to calculate power consumption to ensure power line and transformer voltage and current limits are not exceeded. The relay is also used in the sensing loss of power and the compensating for the loss by communicating to other devices the need to operate switchgears, in order to restore power to the facility as quickly as possible. This type of relay uses technologies that can switch power supplies fast enough that it is barely distinguishable by the naked eye.

### 6.7.1 Relationship between SEL 751A Relay Settings and Circuit Characteristics

According to their paper titled "A Laboratory on the Configuration of Electric Power Substation Monitoring and Control Based on the SEL751A Relay and an Induction Motor Drive for a Three Phase Power Supply", published In the proceedings of the 3rd Interdisciplinary Engineering Design Education Conference, Singh and Wanyama explain that [2 ] *"the relationship between the phase load current $I_j$ and the CT windings (or current through the metering coils of the SEL-751 Relay)$i_j$, is given by Equation 1.*

$$\frac{I_j}{i_j} = \frac{n_j}{N_j} \qquad\qquad 1$$

*Where, $n_j$ is in the numbers of winding turns on the secondary side of the CTs, $N_j$ is the number of winding turns on the primary side of the CTs, and $j$ is the phase identifier. Since a single phase conduct makes up the primary turn of each of the CTs, $N_j = 1$. This reduces Equation 1 to the form given in Equation 2.*

$$I_j = n_j i_j \qquad\qquad 2$$

*On the other hand, the relationship between phase voltages (voltage across the primary windings of the VTs)$V_j$; and the voltage across the input terminals of the SEL-751A (Voltage across the secondary windings of VTs)$v_j$; is given by Equation 3.*

$$\frac{V_j}{v_j} = \frac{N_{vj}}{n_{vj}} \qquad\qquad 3$$

*Where $N_{vj}$ is the number of turns on the primary side of VTs, and $n_{vj}$ in the number of turns on the secondary side of VTs. Letting $n = {N_{vj}}/{n_{vj}}$, reduces the voltage relationship in Equation 3 to the form given in Equation 4.*

$$V_j = n v_j \qquad\qquad 4 \text{ “}$$

The maximum load power requirement of our laboratory is 300W; at 120V (Phase Voltage). The voltage and current levels associated with this load is less than the capacity of the SEL Relay ports. Therefore, there is no need to step the current and the voltage down in order to use the SEL-751A relay to monitor or to control the system. Instead, the load is connected in series with the current measuring coils of the relay. In addition, the load is connected directly to the voltage measuring ports. This results into the circuit shown in Figure 6.7 [2].

According Singh and Wanyama [2], *"The SEL-751A monitors the current $i_j$ and the voltage $v_j$, measuring their peak and root mean square values, as well as the phase angles. Then it uses Equation 2 to determine the current$I_j$, and Equation 4 to determine the load voltage$V_j$. Therefore, for the SEL-751A relay to provide correct information, the value of $n_j$ and of $n$ must be configured into the relay. from Figure 3, we get Equation 5,*

$$I_j = i_j \text{ and } V_j = v_j \qquad\qquad 5$$

*and from Equations 2, 4, and 5, we get Equation 6, which gives the values of $n_j$ and of n for our laboratory.*

$$n_j = n = 1 \qquad\qquad 6"$$

As part of Laboratory 2A, the current and voltage transformer settings described in this section are entered into the relay before configuring the GOOSE message in Laboratory 2B. This is necessary to ensure that the relay records correct current and voltage values.

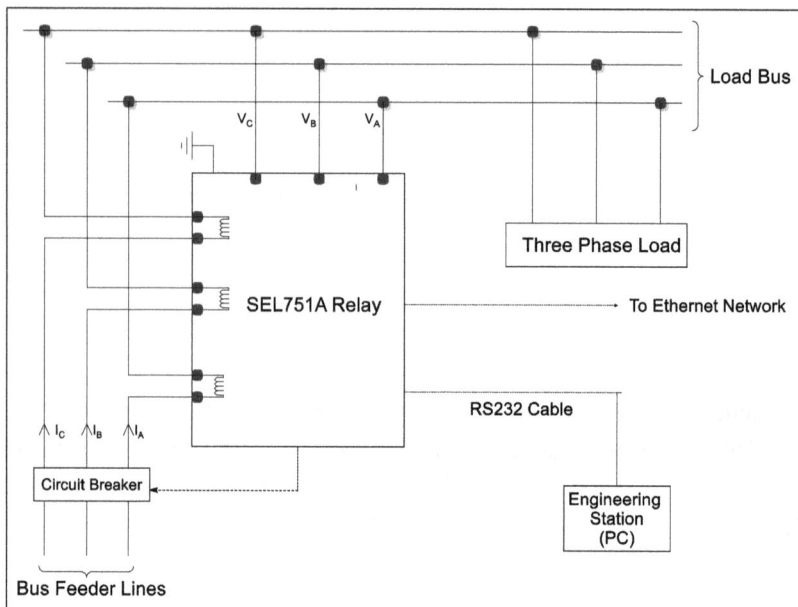

**Figure 6.7: Using SEL-751A Relay without CTs and VTs**

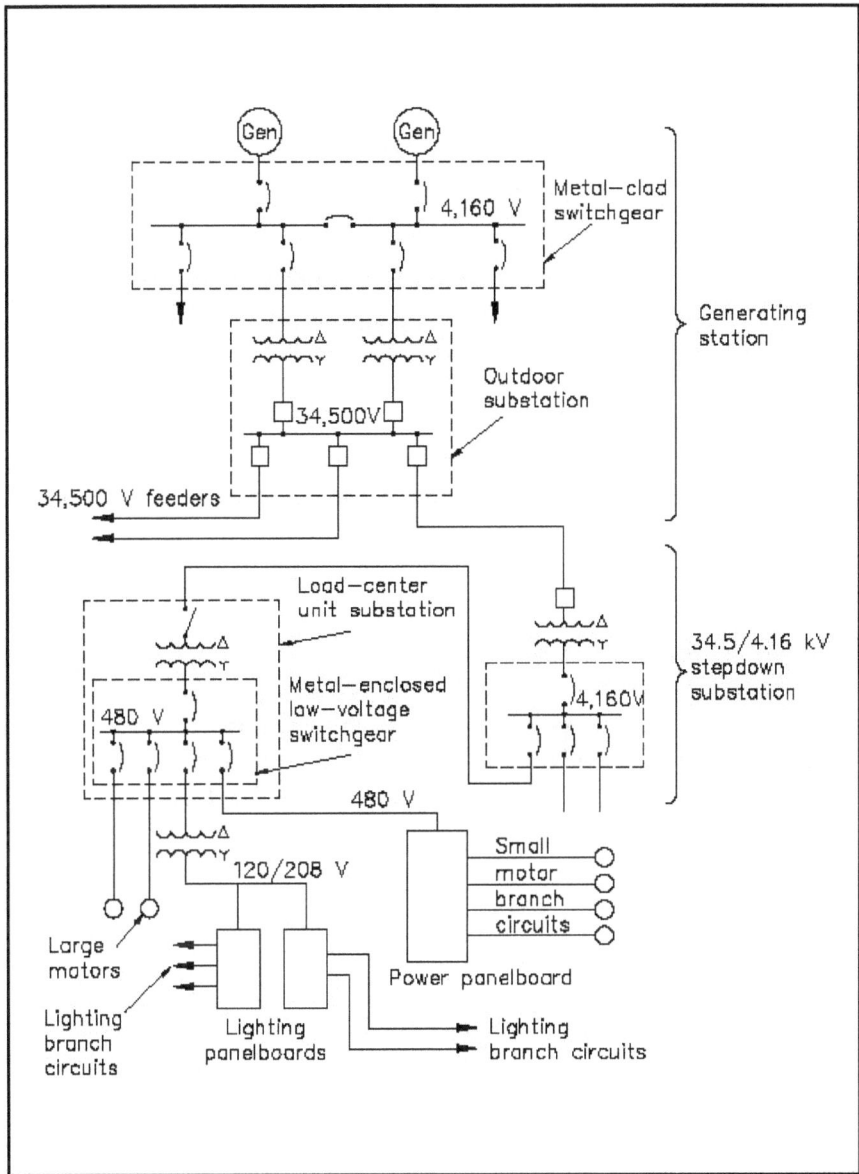

**Figure 6.8: Single-Line Distribution Diagram (Adopted from http://www.tpub.com/doeelecscience/electricalscience2174.htm, on June 13th, 2015).**

[99]

## 6.8    References

[1]  B. Lundqvist, B. Bjorklund & T. Einarsson (n.d.). A user friendly implementation of IEC61850 in a new generation of protection and control devices.       Retrieved       October       2014,       from ABB:http://www05.abb.com/global/scot/scot313.nsf/veritydisplay/3101faf2 df35ad77c125726c00358245/$file/SA2007

[2]  Ishwar Singh and Tom Wanyama, "A Laboratory on the Configuration of Electric Power Substation Monitoring and Control that is based on the SEL751A Relay and an Induction Motor Drive that is used as the three Phase Power Supply", The proceedings of The Interdisciplinary Engineering Design Education Conference (IEDEC), Santa Clara, CA. USA, March 2013.

[3]  Mark Adamiak , Drew Baigent, and Ralph Mackiewicz, IEC 61850 Communication Networks and Systems In Substations: An Overview for Users,     Available     as     of     July     5th,     2015     from http://www.gedigitalenergy.com/multilin/journals/issues/Spring09/IEC6185 0.pdf

[4]  P.T. Manditereza, IEC61850-Enabled Laboratory for Practising Substation Automation and Protection, Available as of June 30, 2015 from Http://aa-rf.org/wp-content/uploads/2014/12/IEC61850-Enabled-Laboratory.docx

[5]  "SCL Manager: Generic Substation State Events (GSSE)". Sclmanager.blogspot.com. Available as of June 1st, 2014

## 6.9    Discussion Questions

**Question 1**

Using a diagram explain the relationship between SEL 751A Relay Settings and Circuit Characteristics

**Question 2**

Compare and contrast IEC61850 and CIP data structures.

**Question 3**

Using IEC81850 network architecture, explain the difference between full and partial implementation of IEC6180 Standard.

**Question 4**

What is the difference between GOOSE and SV messages?

**Question 5**

Using the OSI communication reference model explain how subnetting affects GOOSE messaging?

**Question 6**

Describe how the SEL 751A relay is configured to transmit GOOSE Messages.

**Question 7**

Using an example, describe how AcSelerator QuickSet and AcSelerator Achitect Software applications are used to configure the SEL751A relay response to GOOSE messages.

**Question 8**

Compare and contrast the configuration of IEC61850 devices and Ethernet IP devices.

**Question 9**

How would you modify the subnetworking in Question 3 in Section 3.6 if the machine is Ethernet IP?

## Question 10

How would you modify the subnetworking in Question 3 in Section 3.6 if the network is IEC61850?

## Question 11

Explain your answers for questions 9 and 10.

## Question 12: Project Question

Write the logic that responds to the GOOSE message to transfer Feeder 2 from Circuit Breaker CB-3 to Circuit Breaker CB-1 (Figure 6.9). Assume the two transformer tap changers are already in the same position.

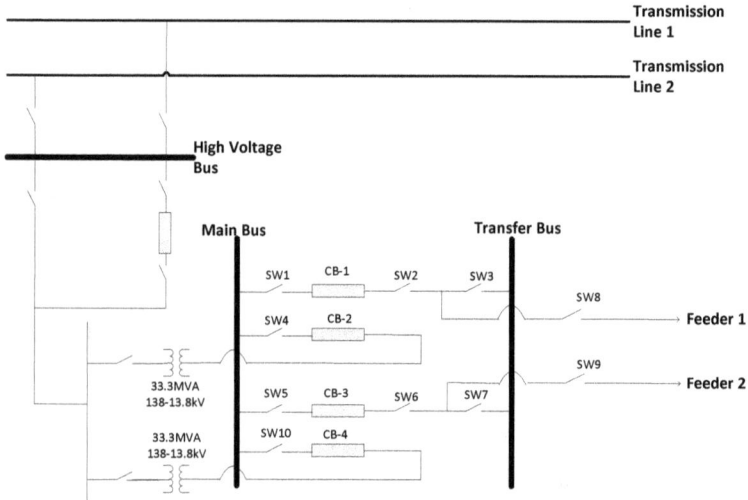

**Figure 6.9: Electrical Distribution Substation**

# Chapter 7  *Integration of Software Applications*

Integrating software applications over the Internet utilizes a multiplicity of web technologies. These technologies are the focus of this chapter.

## 7.1   Historical Background

One of the first solutions to transferring information among software applications was the Microsoft's Windows Clipboard which was based on a simple mechanism of copy and paste. The technology was developed in 1987. With the Clipboard technology, it became possible to create compound documents where a graph created in Excel could be copied and pasted into a word document. However, the relationship between the table that is used to create the graph, and the copied graph in the word document was "static" in the sense that when the data in the table changes, the graph in the word document does not change. To address this issue, Microsoft developed the Object Linking and Embedding (OLE) in 1991. This technology allows a "live" connection between the common objects in primary (source) document and the compound (destination) document. OLE makes it possible to have changes made in the Excel table to be transferred automatically (linked) to the word document and to be edited directly (embedded) there. Another technology, Dynamic Data Exchange (DDE), that had been developed to compete with Windows Clipboard, is now used as the communication protocol of OLE. DDE failed to take off as a technology for creating static compound documents because it is much more complicated than Windows Clipboard. Since OLE technology is based on linking common component objects in applications, it is also referred to as Component Object Model (COM). But it was not until 1995 that OLE started to be referred to as COM, because this is when it was modified to support interactions among other applications other than the text processing ones.

By 1996 networking had taken a hold. Therefore, it became necessary to develop a technology to support interactions among applications located on different computers. This is when Microsoft developed the Distributed Component Object Model (DCOM). But DCOM is not just a modification of COM. Some of its features are rooted in the work of the Open Software Foundation (OSF) whose

[103]

main objective was to make distributed computing possible in heterogeneous environments. OSF developed many standards that are summarized under Distributed Computing Environment (DCE), as well as the Remote Procedure Call technology that is used by DCOM

Technologies that support information transfer among applications are one of the main components of the backbone of the Internet of Things. The focus of this book is not to present material that one can use to develop such technologies, but to present the basic principles of the technologies to enable system integrators to:

- Select software applications that have appropriate information transfer technologies for their projects.

- Identify the type of interfaces that they have to develop in-house in order to integrate their software applications.

The section on web services presents enough material to explain the inner working of the services that one can use it as the initial reference to developing web services. For those who want to learn how to develop these technologies, there are many resources online to refer to.

## 7.2    DCOM

DCOM stands for Distributed Component Object Model and is a set of Microsoft models and program interfaces in which object of client programs can request services from objects of server programs located on another computer on a network. It is based on the Component Object Model (COM), which provides a set of interfaces allowing clients and servers to communicate on the same computer, and on the product of the work of OSF. Imagine a website that contains a script that is processed by another specialized server on the network, other than the one that hosts the website. A DCOM interface can be used by the website program (now acting as a client object) to forward a Remote Procedure Call ( RPC ) to the specialized server object, which provides the necessary processing and returns the result to the website program, which in turn passes the result on to the webpage viewer.

In terms of services, DCOM is generally equivalent to the Common Object Request Broker Architecture (CORBA ). In other words, DCOM is Microsoft's approach to a network-wide environment for program and data objects while

CORBA is sponsored by the rest of the information technology industry under the auspices of the Object Management Group (OMG ).

DCOM is primarily based on windows technology, making it unusable in heterogeneous environments. Moreover, it uses special ports for communications among computers on different networks. This makes the associated network configurations complicated, and in some cases impossible to communicate across firewalls. The other shortfall of using DCOM is that fact that its components can communicate only if they are created using the same programming language. To circumvent the shortfalls of DCOM, CORBA and web services technologies have been developed. CORBA was developed to support communication in heterogeneous environments. Web services on the other hand, utilize the HyperText Transfer Protocol (HTTP) which uses a special port 80, making it easy to send messages across firewalls. Moreover, web service applications are easier than DCOM ones to develop and deploy.

## 7.3    CORBA

CORBA stands for Common Object Request Broker Architecture. It enables communication among software applications written in different languages and running on different computers to work with each other seamlessly. Therefore, CORBA supports communication in a heterogeneous environment. The implementation details from specific operating systems, programming languages, and hardware platforms are all removed from the responsibility of developers who use CORBA. CORBA normalizes the method-call semantics between application objects residing either in the same address-space (application) or in remote address-spaces (same host, or remote host on a network).

CORBA uses Interface Definition Language (IDL) to specify the interfaces, that objects present to the outer world. It then specifies a mapping from IDL to a specific implementation language. Standard mappings exist for Ada, C, C++, C++11, COBOL, Java, Lisp, PL/I, Python, Ruby and Smalltalk. There are also non-standard mappings for C#, Erlang, Perl, Tcl and Visual Basic implemented by object request brokers (ORBs) written for those languages.

The CORBA specification dictates there shall be an ORB through which an application would interact with other objects. This is how it is implemented in practice:

1. The application simply initializes the ORB, and accesses an internal *Object Adapter*, which maintains things like reference counting, object (and reference) instantiation policies, and object lifetime policies.

2. The *Object Adapter* is used to register instances of the *generated code classes*. Generated code classes are the result of compiling the user IDL code, which translates the high-level interface definition into an OS- and language-specific class base for use by the user application. This step is necessary in order to enforce CORBA semantics and provide a clean user process for interfacing with the CORBA infrastructure.

In order to build a system that uses or implements a CORBA-based distributed object interface, a developer must either obtain or write the IDL code that defines the object-oriented interface to the logic the system will use or implement. ORB implementation software usually includes a tool called IDL compiler that translates the IDL interface into the target language for use in that part of the system. A traditional compiler then compiles the generated code to create the linkable-object files for use in the application. Figure 7.1 illustrates how the generated code is used within the CORBA infrastructure. Moreover, The Figure shows the high-level paradigm for remote inter-process communications using CORBA.

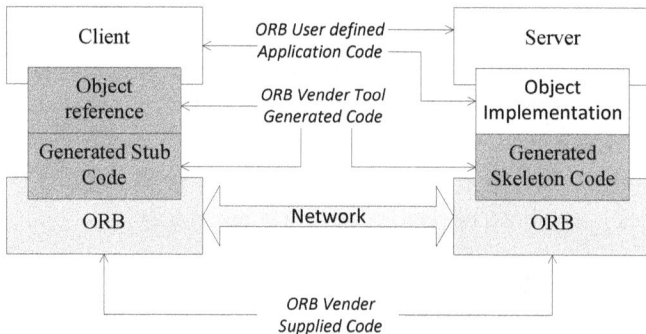

**Figure 7.1: Components of CORBA Code**

Besides providing users with a language and a platform-neutral Remote Procedure Call (RPC) specification, CORBA defines commonly needed services such as transactions and security, events, time, and other domain-specific interface models. Furthermore, CORBA specification addresses many features of distributed computing applications including the following: data typing, exceptions, network protocols, and communication timeouts. On the other hand, the specification (Figure 7.1) leaves some aspects of the distributed system to the application developer to define including object lifetimes (although reference counting semantics are available to applications), redundancy/fail-over, memory management, dynamic load balancing, and application-oriented models such as the separation between display/data/control semantics.

CORBA is not tied to any particular communications transport. Instead it uses raw TCP/IP connections in order to transmit data. Raw TCP/IP bypasses some of the ways the computer handles TCP/IP. Rather than going through the normal layers of encapsulation and decapsulation that the TCP/IP stack on the kernel does, packet is handed to the application that needs it. That is, no TCP/IP processing is done to the packet; it is a raw packet. The application that uses the packet is responsible for stripping off the headers, analyzing the packet, and doing other tasks the TCP/IP stack in the kernel normally does.

A raw socket is a socket that takes packets, bypasses the normal TCP/IP processing, and sends them to the application that wants them. Therefore, if the client is behind a restrictive firewall or transparent proxy server environment that only allows HTTP connections to the outside through port 80, communication may be impossible. Connection can only be possible if the proxy server allows the HTTP connect method or socks connections. In fact same ORBs use random ports making it even more difficult for them to communicate across firewalls. This has forced some distributed systems application developer to switch to using web services.

## 7.4 DDE

**DDE stands for** Dynamic Data Exchange (DDE) and it allows information to be shared or communicated among software applications. For example DDE can be used such that when a data item in a spreadsheet document is changed, it also changes the item anywhere it occurs in other programs. DDE is an Inter-Process Communication (IPC) that uses shared memory as a common exchange area and provides applications with a protocol or set of commands and message formats. DDE uses a client/server model in which the application requesting data is considered the client and the application providing data is considered the server. Many applications use DDE, including Microsoft's Excel, Word, Lotus 1-2-3, AmiPro, Quattro Pro, and Visual Basic. NetDDE is a flavour of DDE that allows programs to converse across networks. For example, a database on one network node could be updated whenever an Excel program in network node was updated. Both nodes must have NetDDE installed.

DCOM provide a "strong binding" between applications because it allows an application to call procedures in other application. On the other hand, DDE is a "weak binding" of applications because it simply supports data exchange through client server capability. DDE client can be implemented in Visual C++ as Windows applications that communicate with other applications or Visual Basic Microinstruction (Macros) that runs within application that share data. To implement DDE applications need access to metadata of the applications that you need to access (for example the row and column of Excel document where data points are stored). Such information is usually not available especially when accessing applications of other organizations. But even when it is available, it is may change without any notification. In cases were metadata is unknown or may change, it is better to use CORBA, DCOM or web services technology.

## 7.5    Web Services

Industrial Internet of Things depends on machine-to-machine communication. Machine-to-machine communication is where machines use network resources to communicate with remote application infrastructure for the purposes of monitoring and controlling themselves, other machines, and their environment. On the other hand Industrial Internet of Things is the interconnection of machines and the physical world to enable them to optimize their processes using information that comes from beyond their primary industries. Moreover, IIoT gives decision makers insight into internal inter-working of the global industrial and business operations of their companies. In fact, IIoT is about using machine-to-machine communication to discover and address industrial needs and challenges. One of the ways machines communicate is through the use of Web Services.

Web services evolved from legacy technologies such as RPC, ORPC (DCOM, CORBA and JAVA RMI). They were developed to solve the following issues:

- Interoperability: Applications based on legacy technologies suffered from interoperability issues because each vendor implements their own proprietary format for distributed object messaging. For example DCOM-based applications work only on Windows operating systems, while RMI applications are to Java programming language.

- Firewall traversal: Distributed applications that use legacy technologies such as CORBA and DCOM use non-standard ports. Therefore, they are usually blocked by firewall servers, hence the need for a technology that relies on HTTP as a transport protocol. Since most firewalls allow HTTP messages access though port 80, using HTTP messages leads to easier and dynamic collaboration.
- Complexity: Most legacy technologies such as RMI, COM, and CORBA involve a whole learning curve.

Web Services are software applications that enable machine-to-machine communication over networks. They are based on open standards such as XML, SOAP, and HTTP. Generally, there are many definitions of the term "Web Services"; some of them are listed here.

- *"A web service is any piece of software that makes itself available over the internet and uses a standardized XML messaging system. XML is used to encode all communications to a web service. For example, a client invokes a web service by sending an XML message, then waits for a corresponding XML response. As all communication is in XML, web services are not tied to any one operating system or programming language--Java can talk with Perl; Windows applications can talk with Unix applications".*
- *"Web services are self-contained, modular, distributed, dynamic applications that can be described, published, located, or invoked over the network to create products, processes, and supply chains. These applications can be local, distributed, or web-based. Web services are built on top of open standards such as TCP/IP, HTTP, Java, HTML, and XML".*
- *"Web services are XML-based information exchange systems that use the Internet for direct application-to-application interaction. These systems can include programs, objects, messages, or documents".*
- *"A web service is a collection of open protocols and standards used for exchanging data between applications or systems. Software applications written in various programming languages and running on various platforms can use web services to exchange data over computer networks like the Internet in a manner similar to inter-process communication on a*

*single computer. This interoperability (e.g., between Java and Python, or Windows and Linux applications) is due to the use of open standards".*

A review of the above definitions of web services revile that they have the following features:

- They are accessible over the Internet or private (intranet) networks, via open protocols (HTTP, SMTP, etc.)
- They use standardized XML messaging system (SOAP)
- They use XML Schema , and they are not tied to any one operating system or programming language
- They are self-describing via a common XML grammar using WSDL
- They are discoverable via a simple find mechanism (UDDI)

### 7.5.1 Components of Web Services

Components of web services can be specified using a protocol stack, and Figure 7.2 shows the web services protocol stack that is still evolving, and has four main layers, namely: discovery, description, messaging, and transport. These layers provide the following functionality:

- The service discovery layer is responsible for populating services into a common registry and providing easy publish/find functionality. Currently, it is managed by Universal Description, Discovery, and Integration (UDDI).
- The service description layer deals with the describing of the public interface to a specific web service. This service is currently handled through the Web Service Description Language (WSDL).
- The service messaging layer is responsible for encoding messages in a common XML format so that messages can be understood at either end. Currently, it is handled by using XML-RPC (Remote Procedure Call) and SOAP.
- The service transport layer is responsible for transporting messages between applications. Currently, this layer uses the following protocols: Hyper Text Transport Protocol (HTTP), Simple Mail Transfer Protocol (SMTP), File Transfer Protocol (FTP), and newer protocols such as Blocks Extensible Exchange Protocol (BEEP).

## 7.5.1.1  UDDI

UDDI stands for Universal Description, Discovery, and Integration. It is an XML-based standard for describing, publishing, and finding web services. It provides a registry for businesses worldwide to list the services they provide on the Internet. Its initial goal was to streamline online transactions by enabling companies to find one another on the Web and make their systems interoperable for e-commerce. Now it is expected to support the Internet of Things. UDDI can be compared to a telephone book such as the white, yellow, or green pages. It allows service providers to list themselves by name, product, location, web services offered, as well as the methods that are called to provide the services. Each of the listings in UDDI is based on WSDL.

| Layer | Standard |
|---|---|
| Discovery | UDDI |
| Description | WSDL |
| Message | XML, SOAP |
| Transport | HTTP, SMTP, FTP, BEEP |

**Figure 7.2: layers of the Web Services Protocol Stack**

## 7.5.1.2  WSDL

WSDL stands for Web Services Description Language. It is an XML document that describes a web service. It specifies the location of the service and the operations (or methods) the service exposes. WSDL became a W3C recommendation in June of 2007. It is independent of the application development platform.

## 7.5.1.3  XML

XML stands for eXtensible Markup Language (XML). It was designed to describe data. XML is a markup language much like HTML, but unlike HTML that is static, XML is extensible because it lets you define your own tags, hence extending the language. Figure 7.3a shows an example of a HTML file, while Figure 7.3b shows a corresponding XML file.

| <html> <body>   <h2>John  Jon</h2>   <p>2 Main Road>    Hamilton<br>    X8X 8X8<br>    john.jon@email.com<br>    </p> </body> </html> | <?xml version=1.0?> <contact>   <name>John joe</name>   <address>2 Main Road</ address>   <city>Hamilton</city>   <postcode>X8X 8X8</postcode>   <email>john.jon@email.com</ email> </contact> |
|:---:|:---:|
| (a) | (b) |

**Figure 7.3: Example HTML and XML Files**

The HTML file displayed the following information in a web browser:

**John Jon**

2 Main Road

Hamilton

X8X 8X8

john.jon@email.com

Basically, HTML specifies "how the document is to be displayed", and not "what information is contained in the document". It is therefore difficult for machines to extract the information that is embedded in HTML file. On the other hand, it is relatively easy for humans to do so. Information contained in the XML file (Figure 3b) is also marked, but not for displaying. It is marked so that machines can extract the content of the file and use it according to their applications and logic. Note that the file is also readable by humans.

XML specifications provide the foundation of all web services technologies. Using the XML syntax that includes instance data, type, structure, and semantic information associated with data to compose messages, allows exchange of data between different programming languages (such as Java, C#, or Visual Basic), middleware, and database management systems. Furthermore, XML is independent of operating systems. This allows it to support web service in the heterogeneous environment of the Web. XML syntax provides the data encapsulation and transmission format for the exchanged data among web services-enabled applications. These applications use the XML syntax to specify how data is represented and transmitted, and how services interact with the referenced applications.

[112]

### 7.5.1.4 SOAP

SOAP stands for *Simple Object Access Protocol. It* specifies how to encode a transport protocol (e.g. HTTP) header and an XML file so that a program in one machine can call a program in another machine and pass along information. It also specifies how the called program can return a response. Generally, SOAP defines the XML-based message format that enable web service applications to communicate and inter-operate with each other over the Web. SOAP provides a standard for formatting web messages, hence enabling information sharing in the heterogeneous environment of the Web. Furthermore, when SOAP messages are transported using HTTP, and they easily get through firewall servers because HTTP is typically Port 80 compliant. On the other hand other messages (e.g. DCOM messages) may be blocked for security reasons. Consequently, applications that use SOAP to communicate can be sure to communicate with other applications programs anywhere.

### 7.5.1.5 HTTP

HTTP stands for HyperText Transfer Protocol. The protocol specifies messages which are formatted and transmitted on the Web. It also defines the actions web servers and browsers take in response to various commands. For example, when a URL is entered in browser, the browser sends an HTTP command to the web server directing it to fetch and transmit the requested web page. Note that HTML defines how web pages are formatted and displayed.

HTTP is currently the most popular option for web services transport, because it is simple, stable, and widely used. In addition, most firewalls allow HTTP traffic, which allows XML-RPC or SOAP messages to masquerade as HTTP messages. While this is good for integrating remote applications, it raises security concerns that must be addressed. Other protocols such as Blocks Extensible Exchange Protocol (BEEP), Simple mail Transfer Protocol (SMTP) and File Transfer Protocol (FTP) can be used to transport SOAP messages (Figure 7.1).

### 7.5.2   How Web Service Work

The web services architecture is based upon the interactions between three roles: service provider, service registry and service requestor. The interactions involve the following operations: publish, find and bind (Figure 7.4). A web service enables communication among various applications using open standards as follows:

- Discover from UDDI the web service provided by an application, including its web address and its method that must be called to get the service. The information about the web service is described based on WSDL.
- Use XML to tag the data. That is, use XML to describe the meaning of the message content.
- Create an HTTP message based on the format described by SOAP.
- Send the HTTP message using TCP/IP.

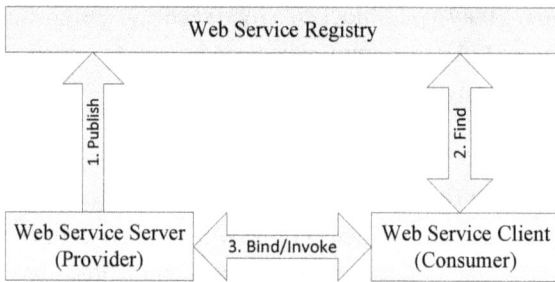

**Figure 7.4: Web Services Architecture**

### 7.5.3    Implementation of Web Services

The .Net environment supports for two approaches for developing web services, namely: the web application (web services) approach and the Windows (Win Forms) application approach. The Figure 7.5 shows that the layer below the forms provides a set of services to the programming process, including base classes and programming support. The Common Language Runtime (CLR) upon which .Net is based, is located above the operated system, and it can be ported to non-Windows operating systems. Moreover, CLR enable all .NET programming languages to access components written in other .Net languages, both at design and run time. The .NET environment generates the intermediate code (MIL – Microsoft Intermediate Language), as well as the application meta-data automatically.

Using the .NET environment to develop web services-enabled applications allows you to take advantage of the inherent communication features of the environment to produce key components of the applications automatically. .NET components

[114]

share information using SOAP (HTTP and XML) or binary coding via TCP. This means that these components can offer firewall friendly communication through HTTP and XML, or high performance communication via binary coding on Intranets. Components that do not support .NET can be accessed using SOAP, while DCOM components can be accessed using binary coding of the data stream.

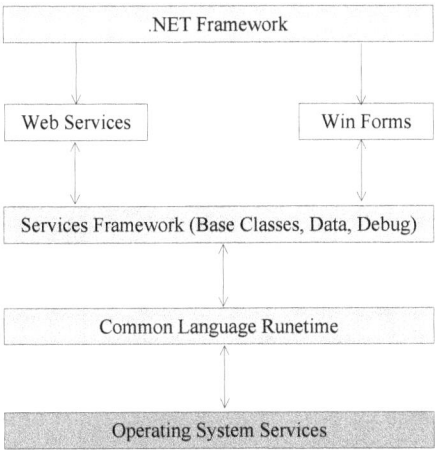

**Figure 7.5: .NET Framework as Used to Develop Web Services**

## 7.5.4 Example: Development of Web Service Application

In this example we develop a service provider that exposes two methods (Distance and Location) as the web services to be used by applications. The "LocationDistance" method returns the distance the shipment is from its destination, and the "LocationName" method returns the name of the location of the shipment. Note that the solution of this example is a standard template for a web service, thus it can be used to develop more sophisticated web service applications. .NET web services use the .asmx extension. The methods exposed as web services have the "WebMethod" attribute, and the file of the web service is saved as myFirstService.asmx in the IIS virtual directory as explained in configuring IIS; for example, c:\MyFirstWebServices (Figure 7.6).

### 7.5.4.1 Web Service Provider

The web service must be published either on an intranet or the Internet in order to test it. In this example, the web service is published on IIS (Internet Information Services) running on a local machine using the following steps:

[115]

- Open Start → Settings → Control Panel → Administrative tools → Internet Services Manager.
- Expand and right-click on the default web site; select New → Virtual Directory. The Virtual Directory Creation Wizard opens. Click Next.
- The "Virtual Directory Alias" screen opens. Type the virtual directory name "MyFirstWebServices" and click Next.
- The "Web Site Content Directory" screen opens.
- Enter the directory path name for the virtual directory "c:\MyFirstWebServices" and Click Next.
- The "Access Permission" screen opens. Change the settings as per your requirements. In this example, keep the default settings.
- Click the Next button. It completes the IIS configuration.
- Click Finish to complete the configuration.

To test whether the IIS has been configured properly, copy any HTML such as unknown.html into the virtual directory (C:\MyFirstWebServices) created above: open Internet Explorer; and type http://localhost/MyFirstWebServices/unknown.html; unknown.html should be displayed in the browser. If it does not work, try replacing the localhost with the IP address of the PC, otherwise, check whether IIS is running. The IIS and the Virtual Directory may need to be reconfigured. Copy myFirstService.asmx into the IIS virtual directory created above (C:\MyFirstWebServices), and open the web service in Internet Explorer (http://localhost/MyWebServices/myFirstService.asmx). It should open your web service page, and the page should have links to the two methods exposed as web services by the application.

```
myFirstService.asmx
<%@ WebService language = "C" class = "myFirstService" %>
using System:
using System.Web.Services:
using System.Xml.Serialization:
[WebService(Namespace="http://localhost/MyFirstWebServices/")]
public class myFirstService : WebService
{
    [WebMethod]
    public int LocationDistance(int d)
    {
      return d:
    }
    [WebMethod]
    public String LocationName(string Lname)
    {
      return Lname:
    }
}
```

**Figure 7.6: Web Service Provider Code**

### 7.5.4.2 *Web Service Based Web Services Consumer*

Let us write a web service consumer using the web services feature of the .NET environment (Figure 7.5). Let us call it myWebApp.aspx (Figure 7.7) and save it in the virtual directory of the web service (c:\MyFirstWebServices\myWebApp.axpx). The file name ends with .axpx because it is for an ASP.NET application. This application has an "Execute" button which when clicked gets the location name and distance from the warehouse.

[117]

| myWebApp.axpx | |
|---|---|
| `<%@ Page Language="C#" %>`<br>`<script runwebservice="server">`<br><br>void runWService_Click(Object sender, EventArgs e){<br><br>myFirstService mySvc = new myFirstService();<br><br>Label1.Text = mySvc.LocationName();<br><br>Label2.Text = mySvc.LocationDistance();<br><br>}<br>`</script>`<br>`<html>`<br>`<head> </head>`<br>`<body>`<br>`<form runwebservice="server">`<br>`<p>`<br>`<strong><u>`Web Service Result -`</u></strong>`<br>`</p>` | `<p>`<br>`<em>`Location Name`</em>` :<br>`<asp:Label id="Label1" runwebservice="server" Font-Underline="True">`Label`</asp:Label>`<br>`</p>`<br>`<p>`<br>`<em>`Location Distance`</em>` :<br>& `<asp:Label id="Label2" runwebservice="server" Font-Underline="True">`Label`</asp:Label>`<br>`</p>`<br>`<p align="left">`<br>`<asp:Button id="runWService"` onclick="runWService_Click" runwebservice ="server" Text="Execute">`</asp:Button>`<br>`</p>`<br><br>`</form>`<br>`</body>`<br>`</html>` |

**Figure 7.7: Web Service Based Web Services Consumer Code**

### 7.5.4.3 Testing the Web Service

Once the web serve provider and consumer are created, create a proxy for the web service to be consumed. Note that this is done automatically by Visual Studio .NET, using the WSDL utility supplied with the .NET SDK. The .Net environment also takes care of the XML message creation and packaging. The WSDL utility extracts information from the web service and creates a proxy. Therefore, the proxy is valid only for the associated web service. For our example, Visual Studio .NET creates myFirstSevice.cs file in the current directory in step 1 below. In the second step we compile myFirstSevice.cs file to create myFirstService.dll (proxy) for the Web Service. An object of the web service proxy is instantiated in the consumer, and the proxy takes care of the interaction between the web service provider and consumer.

1. c:>                                                          WSDL
   http://localhost/MyFirstWebServices/myFirstService.asmx?WSDL
2. c:> csc /t:library myFirstService.cs

3. Put the compiled proxy in the bin directory of the virtual directory of the Web Service (c:\MyFirtsWebServices\bin). Internet Information Services (IIS) looks for the proxy in this directory.

4. Type the URL of the consumer in a web browser to test it (for example, http://localhost/MyFirstWebServices/myWebApp.aspx).

*7.5.4.4 Windows Application Based Web Services Consumer*

Let us now write a web service consumer using the Windows Application feature of the .NET environment (Figure 7.5). Note that this process is the same as writing any other Windows application. We would only need to create a proxy as explained in Section 7.5.4.3, and reference this proxy when compiling the application. Figure 7.8 shows our Windows application that also consumes the web service in Figure 7.6.

| myWinApp.cs |
|---|
| using System.IO;<br><br>namespace SvcConsumer {<br>  class SvcEater {<br><br>    public static void Main(String[] args) {<br>      myFirstService mySvc = new myFirstService();<br>      Console.WriteLine("Calling Location Name Service: " + mySvc.LocationName());<br>      Console.WriteLine("Calling Location Distance Service: " + mySvc.LocationName());<br>    }<br>  }<br>} |

**Figure 7.8: Windows Application Based Web Service Consumer Code**

## 7.5.5 Challenges of Web Services

The challenges of web services are associated with the following three security issues:

- **Confidentiality:** Web services use XML-RPC and SOAP that run primarily on top of HTTP which supports for Secure Sockets Layer (SSL). Therefore web service messages can be encrypted via SSL, a proven and widely deployed technology. However, a single web service may consist of a chain of applications. For example, one large service might tie together the services of three other applications. In this case, SSL is not adequate; the messages need to be encrypted at each node along the service path, and each node represents a potential weak link in the chain. Currently, there is no agreed-upon solution to this issue, but one promising solution is the W3C XML Encryption Standard. This standard provides a framework for encrypting and decrypting entire XML documents or just portions of an XML document. For more information about this standard refer to the W3C website at http://www.w3.org/Encryption

- **Authentication:** There are three options for dealing with the identity of users and user authorizations. Firstly, HTTP has in-built support for Basic and Digest authentication that is used currently to protect HTML documents. This authentication method can be used to protect web services. Secondly, SOAP Digital Signature (SOAP-DSIG) utilizes public key cryptography to digitally sign SOAP messages. This enables clients or servers to validate the identity of the other parties. Thirdly, Security Assertion Markup Language (SAML) is being developed by the Organization for the Advancement of Structured Information Standards (OASIS). Unfortunately there is no consensus on any of these authentication approaches.

- **Network Security:** Generally, there no easy answer to the network security problem. But if you are truly devoted to filtering out SOAP or XML-RPC messages, one option is to filter out all HTTP POST requests that set their content type to text/xml. Another option is to filter the SOAPAction HTTP header attribute. Otherwise, firewall vendors are currently developing tools for filtering web service traffic.

## 7.6    References

[1] Frank Iwanitz and Jurgen Lange, OPC: Fundamentals, Implementation, and Application, Huthig Verlag heidelbery, 3rd Rev. Edition, 2006, ISBN: 3-7785-2904-8

[2] Object management Group, CORBA Basics, Available as of August 27th, 2015 from http://www.omg.org/gettingstarted/corbafaq.htm

[3] W3Schools.Com, Web Services Tutorial, Available as of August 27th, 2015 from http://www.w3schools.com/webservices/

[4] Microsoft: TechNet, Distributed Component Object Model, Available as of August 27th, 2015 from https://technet.microsoft.com/en-us/library/cc958799.aspx

## 7.7    Discussion Questions

**Question 1**

Describe the history of DCOM.

**Question 2**

What are the performance limitations of DCOM with respect to CORBA?

**Question 3**

What are the performance limitations of DCOM with respect to Web Services technology?

**Question 4**

What are the performance limitations of CORBA with respect to Web Services technology?

**Question 5**

What are the performance limitations of CORBA with respect to Web Services technology?

**Question 6**

Explain the advantages of using the .NET environment to develop web services applications.

**Question 7**

Explain how the .NET environment is used to develop web services applications.

**Question 8 (Project Question)**

i)   Write a web service application that take six integers for the consumer and uses the provider to calculate the average of the integers. Assume the consumer is on different computer that is located on the same network as the provider's computer.

ii)  Explain how you would compile and deploy the web service.

iii) Explain how your code would change if the consumer was on another network, behind a firewall.

# Chapter 8  *OPC: Automation Systems Data Access and Integration*

Formerly "OLE for Process Control", OPC stands Open Productivity and Control because its application has been expanded beyond the process control area for which it was first developed. It is an industry standard managed by the OPC Foundation (http://www.opcfoundation.org/) specifying software interfaces (objects, methods) to servers that collect data produced by field devices and programmable logic controllers (Figure 8.2). The main and most applied specifications of OPC are: OPC Data Access, OPC Alarms and Events, and OPC Historical Data Access.

## 8.1    OPC Standard

The original standard was developed in 1996 by an industrial automation industry task force under the name OLE for Process Control (Object Linking and Embedding for Process Control). OPC supports the collection and integration of real-time plant data of control devices from different manufacturers. OPC Foundation (http://www.opcfoundation.org/) was created to maintain the OPC standard and it later renamed the standard Open Platform Communications. This was done because applications of OPC had gone beyond process control to include other areas such discrete manufacturing, building automation, and electrical substation automation. Moreover, OPC has also grown beyond its original OLE (Object Linking and Embedding) implementation to include other data transportation technologies including XML, Microsoft's .NET Framework, and OPC Foundation's binary-encoded TCP format.

No company "owns" OPC, thus anyone can develop an OPC server whether they are members of the OPC Foundation or not. However, it is important to note that since some OPC specifications are available only to members of the OPC Foundation, non-members may not be using the latest specifications. Furthermore, there is no pre-requisite for the system integrators to belong to the foundation, thus it is up to companies that require OPC products to ensure that the products they purchase are certified and that their system integrators have the necessary training.

## 8.2 OPC System Architecture

Before OPC was developed, every device vendor had to develop software for accessing data from their devices in such a way that it interfaced with existing application software. In addition, application developers also had to ensure that their applications were able to connect to the various proprietary interfaces for the drivers that access plant data from automation devices (Figure 8.1). That is, traditionally, any time an application needed access to data from a device, a custom interface, or driver, had to be written.

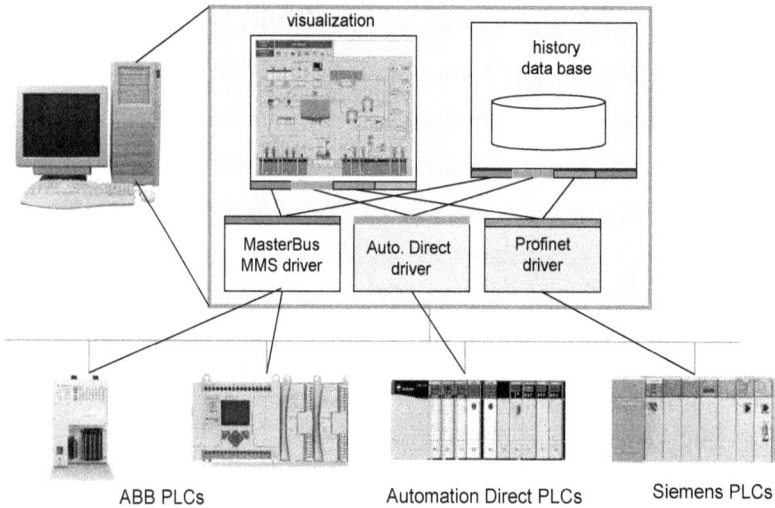

**Figure 8.1: Before OPC**

OPC provides a common bridge between application software and process control hardware drivers. It defines consistent methods of accessing field data from plant floor devices, and these methods are independent of the type and source of data. That is, all OPC servers provide the same methods for OPC clients to access their data. This means that OPC standards define the interface between servers and clients (Figure 8.2).

One of the main objectives of OPC is to reduce the amount of duplicated effort required from hardware manufacturers and their software partners, to interface hardware with SCADA and HMI systems. Once a hardware manufacturer develops an OPC server for a new hardware device, their device can be accessed by any 'top end' application. On the other hand, once a SCADA producer develops an OPC client, that client can access any hardware existing or yet to be created, so long as that device has OPC compliant server.

## 8.3    OPC Data Access (OPC DA)

Process variables which are generated by sensors or calculated in the Programmable Logic Controllers (PLCs) describe the state of the plant. These variables can be sent to OPC servers upon a change, on demand or when a given time elapses. OPC DA specification addresses the collection of process variables. The main clients of OPC DA are visualization and soft-control systems (Hard Control is done by the primary controller). Figure 6.3 shows that OPC tag information is generated by the controller configuration software such

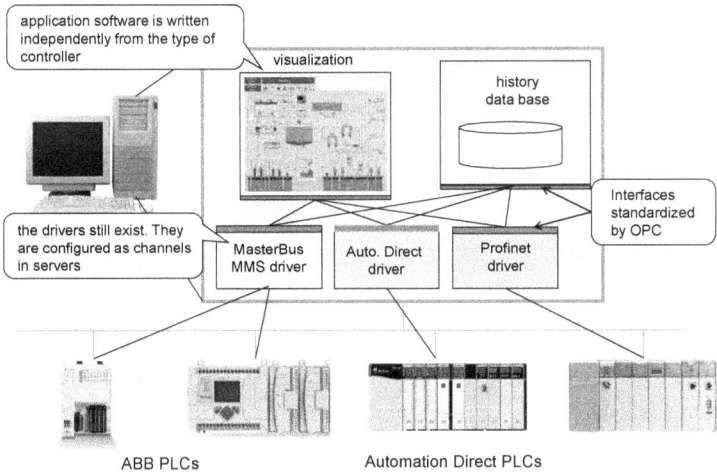

Figure 8.2: OPC Data Interface

RsLogix5000. Then this information is used to configure OPC DA servers. Note that OPC DA servers do not send tag values directly to clients. Instead they send to the clients the memory location of the tags and the clients reads the tag values from that location themselves.

Data in an OPC server is structured as a directory with root, stem, branches and leaves (Figure 8.3). A combination of root, branch and leaf forms items. Items are identified by their "fully qualified ItemID", for example "Process_Line_1.Controller_2.Level_2" shown in Figure 8.3. Branches may contain other branches and items. This structure is usually defined during

[125]

engineering of the attached devices, but intelligent servers could configure themselves by reading the attached devices of a file created during the configuration of devices.

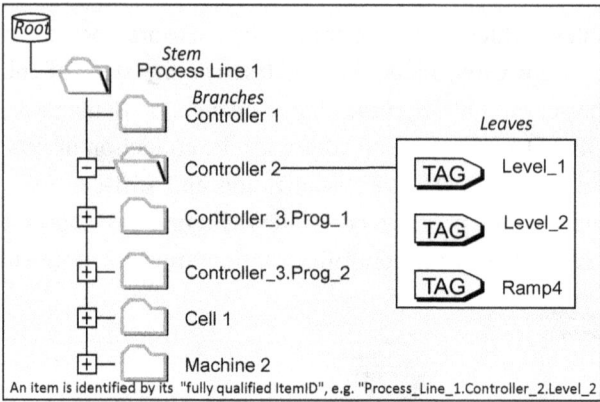

Figure 8.3: Structure of Data in OPC Server

## 8.4    OPC Alarms and Events (OPC AE)

Events are changes in the process that need to be logged, such as "production start", and alarms are abnormal states in the process that require attention, such as "low oil pressure. OPC AE (Alarms and Events) specifies how alarms and events are subscribed, under which conditions they are filtered and sent with their associated messages. The main clients of OPC AE are the Alarms and Event loggers. Alarms and events values are sent by servers to clients. In addition, alarms have to be acknowledged.

An OPC AE Server is configured using the information that comes from the development tools for controllers e.g. PLC. Controllers generate events in response to changes in the plant variables, together with their precise time of occurrence, type, severity and associated message for the human operator. OPC AE servers register these events and make them available to clients. Alarm messages are usually more detailed than events, that they usually require acknowledgement.

The OPC Alarms & Events Interface gives access to the AE server, allowing to:

- browse the OPC AE Server for predefined events.
- enable or disable alarms and events
- subscribe to alarms and events of interest

[126]

- receive the event and alarm notifications with the associated attributes
- acknowledge alarms

OPC AE data is communicated using the "message passing" paradigm, contrarily to OPC DA data that is communicated using the "shared memory" paradigm. In the shared memory paradigm, the server communicated to the client the memory location of the data, and then the client reads the data from the memory. On the other hand using message passing means that an event is kept in a queue until all clients have read it (or timed out). The AE server guarantees that different clients will receive all events in the same sequence.

OPC AE defines three types of events, namely:
- simple: process control system related events (change of a boolean variable)
- condition-related: notifies a change of an alarm condition (CLEARED, ACKNOWLEDGED),
- tracking-related: origin outside of the process (e.g. operator intervention)

Events have to be time stamped, and this can be done at the following locations:
- The device that originally produced the data (external event - low-level event)allowing Sequence-Of-Events with a high accuracy, down to microseconds
- The controller, (internal event) using the controller's clock to time-stamp messages giving accuracy not greater than the period of the tasks, about 1 ms.
- The OPC Server, when an event message arrives (tracking events) not more accurate than DA, about 10 ms)

It is always important to know the state of an alarm, or alarm condition. An alarm condition is a named state machine that describes the state of an alarm. It is define by the following three variables:
- Enabled: the condition is allowed to send event notifications
- Active: the alarm signal is true
- Acknowledged: the alarm has been acknowledged

Alarms and Event are organized by area, and the areas may contain other areas. Unlike branches in OPC DA, areas and sources have properties that allow the user to disable or enable events or alarms by area or by source. Areas and sources are logical locations that usually correspond to parts of the plants, rooms, or specific plant equipment.

## 8.5    OPC Historical Data Access (OPC HDA)

OPC Historical Data Access specifies data log and the main client of OPC HDA are historians and trend loggers. Other features of the OPC such as Dynamic Data Exchange (DDE), Data Bridging, and Tunnelling are discussed in Section 8.4 with respect to the Laboratory on Cogent DataHub OPC client.

## 8.6    Servers and Clients Configuration

Traditionally, servers and clients communicate using the Component Object Mode (COM), and the Distributed Component Object Mode (DCOM) paradigms. While both paradigms allow clients to call procedures in servers, COM works only when the server and the client are located on the same computer. If the server and the client are located on different computers, DCOM is used. It allows data to be sent to a stub in the remote computer from where the client can access it (Figure 8.4).

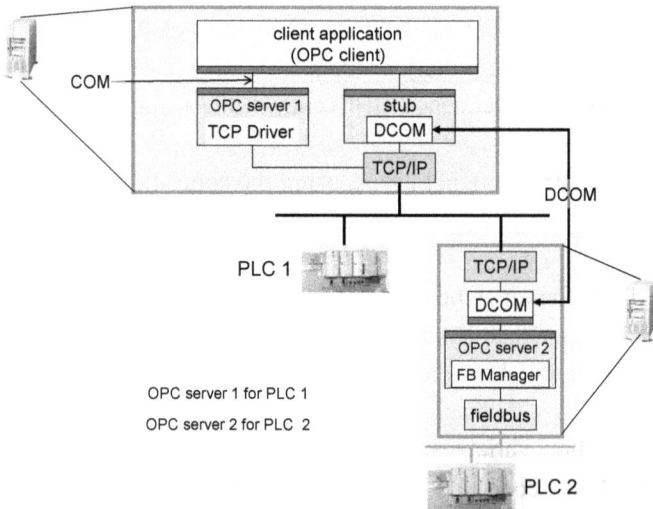

**Figure 8.4: OPC Server and Client on same PC (COM) or on different PCs that are on same Intranet (DCOM)**

If the server and the client are located on different network, the associated network configurations are very complicated; and if the computers are not on the same Intranet (behind firewall) communication cannot be achieved. Furthermore

DCOM only works with Windows computers. Therefore, in such cases it is better to use OPC servers and clients that communicate using web services technology, as known as OPC UA.

## 8.7 OPC UA

There many technologies that support sharing of data among applications (Chapter 7), with varying capabilities. If an OPC server or client is developed using any of these technologies; it inherits there capabilities. Since OPC DA, OPC HDA, and OPC AE are traditionally based on COM/DCOM, they are limited to Windows computers that are on the same Intranet. In order to make OPC useable in a heterogeneous environment such as the Internet which is littered with firewalls, OPC Foundation developed the OPC Unified Architecture (UA) standard which is based on web services. OPC UA supports all the functionalities of the traditional OPC DA, HDA, and AE. It also supports communications of OPC applications that are developed using COM/DCOM technologies. However, OPC UA inherits the security issues of web services (Chapter 7). And OPC UA applications (Clients and Servers) handle these issues in a proprietary manor. Therefore when purchasing OPC UA applications, it is important to take into account how they address security.

## 8.8 OPC Server Laboratory

Laboratory E (Appendix E) focuses on configuring an OPC server (KEPServer) to access manufacturing industry data, as well as electrical substation automation data over Ethernet. Manufacturing data from the PLC is access by the server through Ethernet IP while the electrical substation automation data is accessed through IEC61850 (MMS) protocol (Figure 8.5). While IEC61850 data comes predefined in the IED, manufacturing control data is created through the process of programing PLCs in form of tags. The tags are download on the PLCs together with the control logic, from where it is accessed by the OPC server (Figure 8.6). Once all this data is in the OPC server, it is accessed by OPC clients independent of its source, as well as communication protocol used to access it. This is how OPC and Ethernet technologies are able to integrate industries, leading to the phenomenon referred to in literature as Industrial Internet of Things (IIoT).

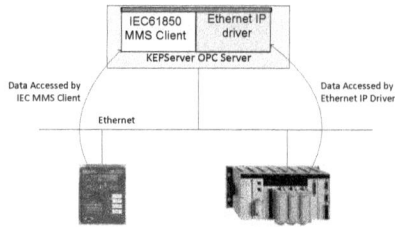

**Figure 8.5: Access of manufacturing and Electrical Substation Data**

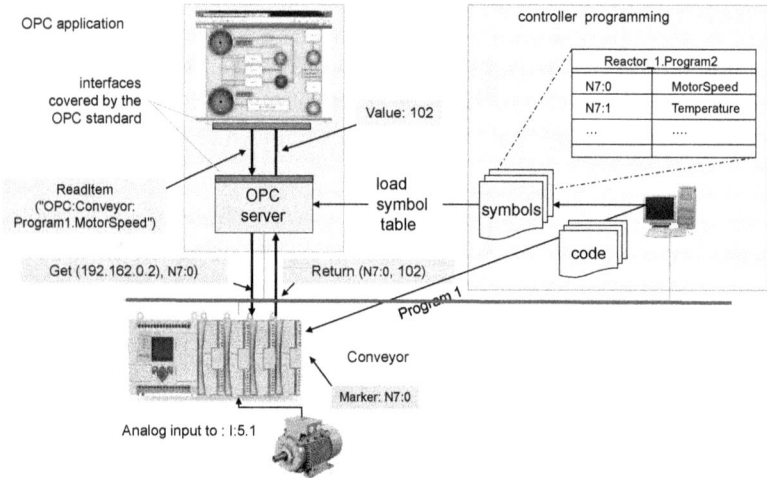

**Figure 8.6: OPC DA**

## 8.8.1 KEPServer OPC Server

KEPServer, made by Kepware Technologies Inc. is one of the most advanced OPC servers. It provides multiple software drivers for accessing data from automation devices used in diverse array of industries, including manufacturing automation, process automation, building automation, electrical smart grid and electrical substation automation, oil and gas, and water and waste water management (Figure 8.5). Moreover, Figure 8.7 shows that KEPServer supports many OPC specifications for client interfaces, such as Alarms and Events (OPC AE), Data Access (OPC DA), Unified Architecture (OPC UA), and Dynamic Data Exchange (DDE) [2].

[130]

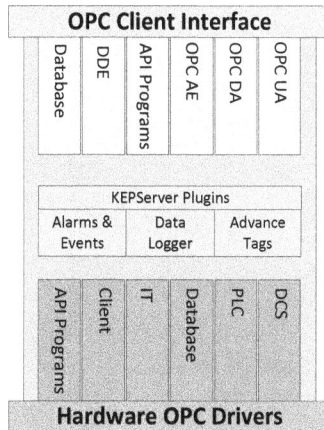

**Figure 8.7: Features of KEPServer OPC Server**

KEPServer OPC Server is designed to allow quick and easy configuration of the communication between the server and automation devices. In fact, Figure 8.8 shows that this process involved the following three main steps:

- Select a driver to create channel: This attaches the software driver that is used to connect the server to the device of your project.
- Specify the device or system to communicate with: The step sets the communication parameters of the channel based on the communication protocol used.
- Select the items or tags for your database: This can be done manually, by reading tags from the device, or by reading a special file created by the device configuration software.

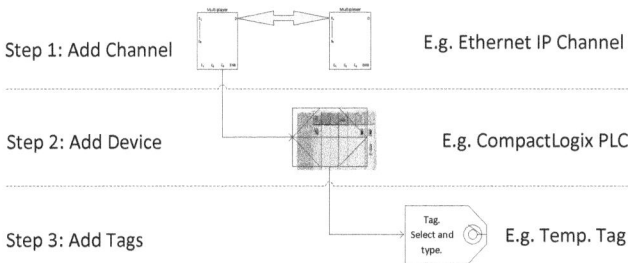

**Figure 8.8: Steps for Configuring KEPServer OPC Server**

## 8.9    OPC Client Laboratory

KEPServer, the OPC server, is configured in Laboratory 3A to access data from a PLC and from a SEL 751A relay, and make it available to OPC clients used to implement HMIs, historians, and alarms and events systems. OPC DataHub, used as the OPC client in Laboratory F (Appendix F) is a powerful OPC application developed by Cogent Real-Time Systems of Georgetown, Ontario. DataHub is based on multiple enabling technologies to supports the following functions [1]:

- OPC Data Access: This is the access to process data in controllers for the purpose of providing HMIs

- OPC Tunneling: means networking data between an OPC server on one computer and an OPC client on another computer, without using the troublesome DCOM networking offered by most OPC vendors.

- OPC Data Logging: The OPC DataHub enables any OPC server to write data to or read data from a relational database such as Access, MS SQL Server, Oracle or any other ODBC compliant database

- OPC Bridging: means connecting an OPC server to another OPC Server on the same computer.

- OPC to Excel: transfer OPC data into Microsoft Excel

- OPC Scripting:  a powerful built-in programming language called Gamma is built in the DataHub

- OPC to e-Mail/SMS: send email and SMS text messages whenever an alarm, or specified timer event occurs.

- OPC Aggregation: means collecting data from several OPC servers into one common point of access.

- OPC to Web

- OPC to MES/ERP

- OPC to LINUX/QNX

### 8.9.1    Major OPC DataHub Specifications

OPC DataHub has the following features:

- Supports OPC Server and Client connections.

- DataHub will connect to OPC DA 3.0 servers (and 2.05a servers that support browsing).

- DataHub will also accept connections from OPC DA 3.0 or 2.05a clients.

- Supports DDE Server and Client connections.

- Supports live data in a web browser using ASP, AJAX and Java.

- Supports custom TCP/IP connections through Java, .NET and C++ DataHub APIs.

- Supports Windows GUI development through built-in Scripting language.

- Supports ODBC compliant database access.

- DataHub supports communication with MATLAB applications.

### 8.9.2    OPC Server Connection Setup

Figure 8.9 shows the main window of the OPC DataHub client. It is through this window that the client is setup to act as an OPC server, an OPC client, or both simultaneously.

In order for OPC DataHub to acts as client, Check the "Act as an OPC Client" box, and in order for it to act as a server to other clients, check the "Act as on OPC Server" box. DataHub can be a client to more than one OPC server. All you need to do is to specify information for each OPC server connection. Once you have a server listed, you can activate or deactivate the connection using its "On" check box in the "OPC DA" area. To add a server, click on the "Add" in the OPC area. The "Define OPC Server" window in Figure 8.10 opens. To edit an already added server, double-click it or select it and click on "Edit" to open the "Define OPC Server" window. To remove a server, highlight it and click the "Remove" button. Clicking the "Reload Data from All Servers" button causes the DataHub to disconnect from all OPC servers, and then reconnect and refresh the data set for each server.

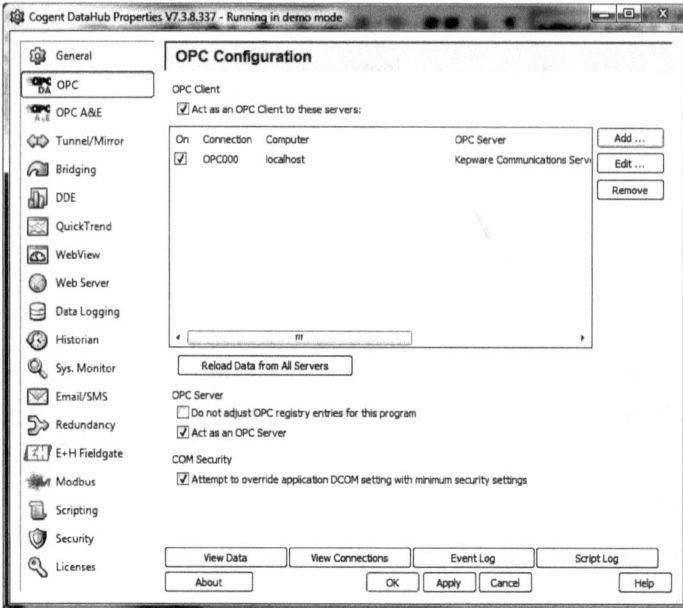

Figure 8.9: Main Window of OPC DataHub Showing its Capability Areas

## 8.9.3    Definition of OPC Server Window

In order for DataHub to access data from an OPC server, the server connection must be defined through the "**Define OPC Server**" window shown in Figure 8.10. The window is opened by clicking the "**Add**" or "**Edit**" button. You can access OPC servers on your local computer by clicking on the dropdown icon on the "OPC Server Name" box. Select the server you want to connect to the client by double clicking on it in the dropdown window (Figure 8.10: KEPServer Ver 5.1 is selected). Moreover, configure the following server connection parameters [1]:

**Connection Name**: A name used by the OPC DataHub to identify the connection. There should be no spaces in the name. It doesn't matter what name is chosen, but it should be unique to other connection names.

**Computer Name**: The name or IP address of the computer running the OPC server you want to connect to. Select it from the drop-down list, or type it in. If the server is on the local PC, refer to it as "localhost".

**OPC Server Name:** The name of the OPC server that you are connecting to, selected from the list of available servers.

[134]

**Data Domain Name:** The name of the DataHub domain in which the data points are received.

**Maximum update rate (milliseconds):** This option lets you specify an update rate, useful for slowing down the rate of incoming data. The default is 0, which causes values to be updated as soon as possible.

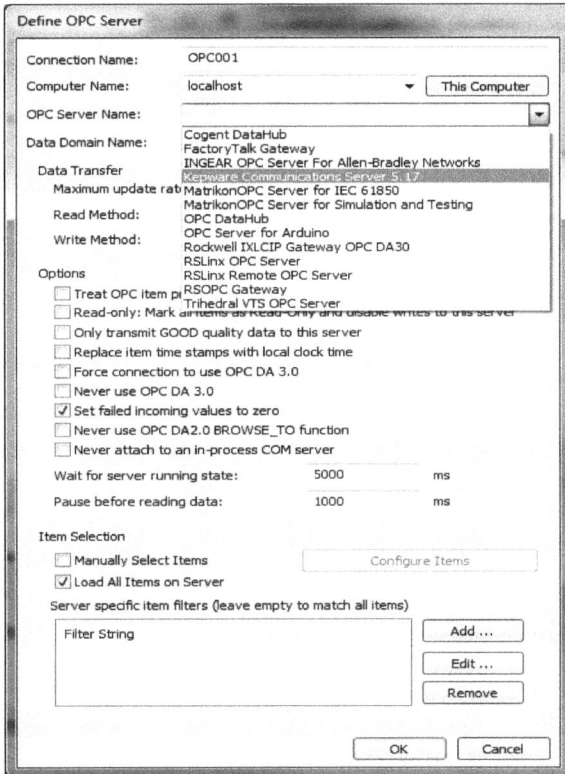

**Figure 8.10: Define OPC Server Window**

**Read Method:** Choose how to read data from the OPC server from one of the following methods [1]:

- **Asynchronous Advise:** The OPC server sends a configured point's data to the DataHub immediately whenever the point changes value. This is the most efficient option, and has the least latency. This is the read method used in Laboratory 3B.

- **Asynchronous Read:** The DataHub polls the OPC server for all configured points on a timed interval (set by the Maximum update rate). This option is less efficient than Asynchronous Advise, and has higher latency.
- **Synchronous Cache Read:** The DataHub polls the OPC server for all configured points on a timed interval (set by the Maximum update rate), and this thread waits for a reply. This option is less efficient than Asynchronous Advise or Read, and has higher latency than either of them.
- **Synchronous Device Read:** The DataHub polls the PLC or other hardware device connected to the OPC server for all configured points on a timed interval (set by the Maximum update rate), and this thread waits for a reply. This is the least efficient of all of these options, and has the highest latency.

**Write Method:** Choose how to write data to the OPC server using one of the following methods:

- **Asynchronous Write:** Provides higher performance. OPC DataHub writes changes in point values to the OPC server without waiting for a response. This is the write method used in Laboratory 3B.
- **Synchronous Write:** Elicits a quicker response from the OPC server, but results in lower overall performance. OPC DataHub writes changes in point values to the OPC server and waits for a response. This option is useful if the OPC server doesn't support asynchronous writes at all, or if it can't handle a large number of them.

**Treat OPC item properties as DataHub points where possible:** This option lets you register and use each OPC item property as point in the DataHub. Some OPC servers are slow to register their OPC items and properties. Using this option with one of these servers can significantly slow the start-up time of the DataHub

**Read only:** Mark all items as Read-Only and disable writes to this server: Here you can specify that the OPC server be read-only, regardless of how individual items are specified. Items in the DataHub that originate from such an OPC server will be read-only to all DataHub clients.

**Replace item time stamps with local clock time:** This option allows you to set the timestamps for the items from this server to local clock time.

[136]

**Force connection to use OPC DA 3.0:** This setting will allow you to choose the DA 3.0 write methods from the "Write Method" drop-down box. It will also instruct the OPC DataHub to attempt to browse the server using DA 3.0 browsing. This setting will override any automatic information that the OPC DataHub may determine about the server based on the server's registry entries.

**Never use OPC DA 3.0:** This setting will remove the DA 3.0 write methods from the "Write Method: drop-down box, and will instruct the OPC DataHub to only use DA 2.0 browsing. This setting will override any automatic information that OPC DataHub may determine about the server based on the server's registry entries.

**Set failed incoming values to zero:** The OPC specification requires an OPC server to send an EMPTY (zero) value whenever it sends a failure code in response to an item change or a read request. Some OPC servers, however, send a valid value with the failure code under certain circumstances. To ignore any such value from the OPC server and assume EMPTY, keep this box checked (the default). If instead you want to use the value supplied by your OPC server, uncheck this box.

**Never Use OPC DA 2.0 BROWSE_TO Function:** This setting will disallow the BROWSE_TO function when communicating with OPC DA 2 servers. Sometimes an OPC server will have problems with this function that prevent OPC DataHub from connecting to it. Checking this box might allow the connection to be established in those cases.

**Manually Select Items:** Selecting this option and clicking the "Configure Items" button opens the OPC Item Selection window, where you can specify exactly which points you wish to use

**Load All Items on Server:** This selection tells the DataHub to register all the points in the OPC server. In the "Server specific item filters" area you can enter one or more strings to filter for groups of items in the OPC server. It only works when "Load All Items on Server" is selected. Use the "Add" or "Edit" button to open the "Edit a filter string" window:

**COM Security**: If connecting the OPC DataHub over a network using tunneling fails; check the box under "COM Security" in the OPC area to relax COM security. This setting will override the COM permission settings for the application, but will not override the system's global COM restrictions. It is common for OPC servers to operate at minimal DCOM security settings, since high security interferes with connectivity and most control systems do not operate in hostile network environments. If in doubt, consult your system administrator.

## 8.10    Data Bridging

OPC DataHub supports OPC server to OPC server connectivity by bridging data from different OPC servers or from the same server but different devices. Figure 8.11 shows bridging of data from a MMS client OPC server on one hand, and data from a manufacturing automation OPC server on the other hand. Data from a MMS client OPC server is tunneled to the client on PC 2 before enabling the data bridge.

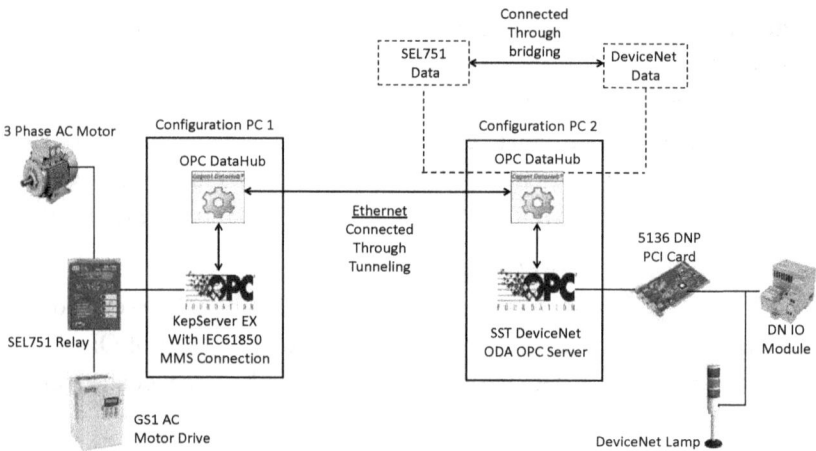

**Figure 8.11: OPC DataHub Data Tunneling and Bridging**

In Laboratory 3B, we use the SEL 751A relay to monitor the power supply of a process automation system controlled by a MicroLogix 1400 PLC. KEPServerEx V5.1 is used to access data from the SEL relay and the PLC using two channels, namely: IEC 61850 MMS channel for the SEL relay, and manufacturing automation Ethernet IP channel for the MicroLogix 1400 PLC. The Server makes the data from both channels available to OPC DataHub client, making it possible to bridge the data as shown in Figure 6.12. The bridging process is covered in detail in laboratory 3B (Appendix F).

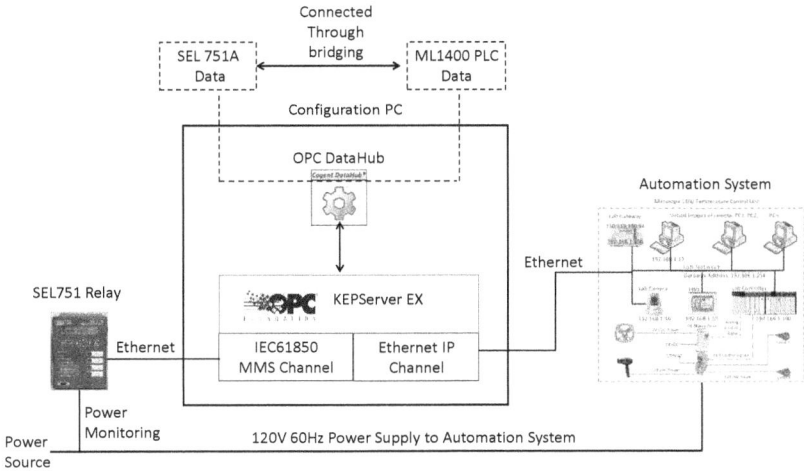

Figure 8.12: OPC DataHub Data Bridging System

## 8.11    OPC as a Data Hub of IIoT

OPC provides a convenient way to exchange data among applications, databases, and controllers. Figure 8.13 shows the following data sharing functionalities of OPC:

- OPC enables sharing of information among controller that use different communication protocols, including controllers that support different value chain participants. For example, OPC enables communication between the CompactLogix PLC that is designed for the manufacturing industry and the SEL751A relay designed for the electricity industry.
- OPC can easily read and write to any Microsoft Office application, as well as other applications.
- OPC applications can share data with many other elements of IIoT such as databases that hold logistics data, government regulations, and environmental information.
- OPC applications can communicate with process, plant and enterprise-wide simulation systems, as well as advanced control systems. Advance control systems can be used to control that plant directly or can be used for supervisory control through OPC enabled SCADA system.

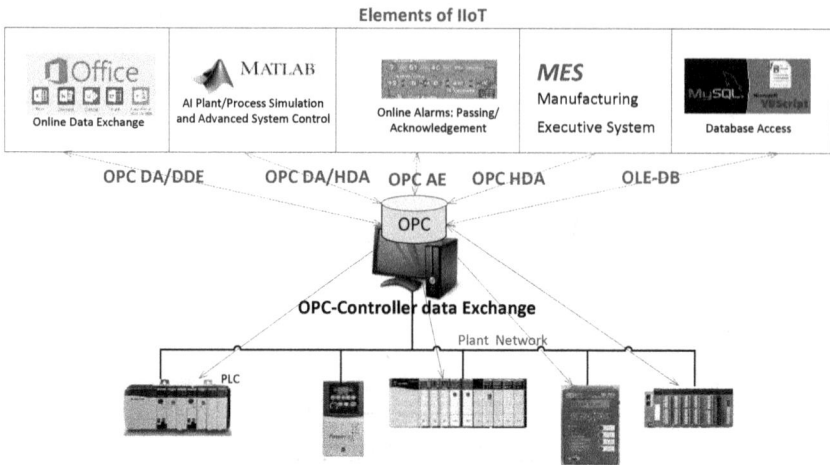

**Figure 8.13: OPC as a Data Hub of IIoT**

By providing the above functionalities, OPC serves as a data hub for IIoT, as well as a data exchange system for the enterprise on one hand, and for the plant on the other. Consequently, OPC is the glue that binds the IIoT together.

## 8.12   References

[1]   Cogent Real-Time Systems, Cogent Documentation, Available as of June 30, 2013   http://www.opcdatahub.com/Docs/cogentdocs.html

[2]   Kepware Technologies Inc. KEPServer Manual, Available as of June 30, 2015                                                              from https://www.kepware.com/products/kepserverex/documents/kepserverex-manual/

[3]   H. Kirrmann, OPC (Open Process Control formerly OLE for Process Control), Lecture slides, Swiss Federal Institute of Technology in Lausanne

## 8.13    Discussion Questions

### Question 1

Using diagrams showing legacy and OPC approaches to data access, and history data access explain the advantages and disadvantages of OPC technology.

### Question 2

How is the KEPServer OPC server able to access data from devices of various industries, such as manufacturing industry PLCs, and electricity industry IEC61850 relays and other control devices?

### Question 3

Using the OSI communication reference model, Allen Bradley ControLogix PLC (CIP), SEL751A relay (MMS), KEPServer OPC Server, and Cogent DataHub OPC Client, explain how Ethernet supports accessing data of devices from multiple industries.

### Question 4

Using the OSI communication reference model, explain the following phrase: "KEPServer OPC server cannot directly access GOOSE related data, but it can access it through MMS".

### Question 5

How is AcSelerator QuickSet used to configure the data that is accessed by OPC servers?

### Question 6

How does KEPServer use AcSelerator QuickSet configuration to populate the MMS tags of SEL751A relay?

### Question 7

Using SEL 751A and MicroLogix 1400 PLC, describe the three methods used to add tags to the KEPServer OPC server.

## Question 8

Describe the communication between KEPServer OPC server and Cogent DataHub OPC client to support the following OPC functions: OPC DA, OPA AE, and OPC HDA.

## Question 9

**Project Question:** Using SEL 751A, MicroLogix 1400 PLC, Arduino Microcontroller, and Raspberry Pie explain how Ethernet, KEPServer, and Cogent DataHub OPC client is used to support Internet of Things

# Chapter 9 *Industrial Internet of Things and Industry 4.0*

This chapter describes how network technologies are used to create Industrial Internet of Things (IIoT), and how IIoT supports Industry 4.0. Moreover, the chapter presents a set of Industry 4.0 examples that explain the roles of the various components of IIoT.

## 9.1    Background to IIoT and Industry 4.0

In 2012 the government of Germany established an Industry 4.0 working group to spur a potential fourth industrial revolution, following the previous three industrial revolutions, namely: the introduction of the steam engine, electricity, and information technology. The main objective of this revolution is to develop new business models that tap the potential optimization in production and logistics caused by increased industrial automation, intelligent system monitoring, autonomous decision-making, and real-time or almost real-time communication at all levels. Similar strategies have been proposed by other main industrial countries. For example, the United States of America through the industrial Internet Consortium has proposed Industrial Internet of Things (IIoT) which is defined in literature as the use of Internet of Things (IoT) technologies in manufacturing. Also known as the Industrial Internet, IIoT incorporates machine learning and big data technology, harnessing the sensor data, machine-to-machine (M2M) communication and automation technologies that have existed in industrial settings for years. The Government of China has proposed an industrial transformation strategy called Internet Plus, which aims to integrate mobile networks, cloud computing, big data, the Internet of things, and other related IT technologies with modern manufacturing. The integration is expected to promote the healthy development of e-commerce, industrial networks and Internet finance, and subsequently increase Chinese Internet companies' presence in the global market. The following three main conclusions can be drawn from the Germany, USA and China strategies to industrialization in the information age:

- The Germany and China strategies are spearheaded by government while the USA strategy is spearheaded by the private sectors. Both approaches have strengths and weaknesses of being affected by general characteristics of government or private sector programs. For example, government programs tend to be thorough but inefficient, while private sector programs tend to be efficient but focused on making money; which may leave out program components that are not expected to make money, but may be needed to strengthen the program for the great good.

- The USA and Chain strategies put more focus on the technologies that are expected to transform industry, while the Germany strategy puts more focus on the use of the technologies to transform industry.
- All three industrial transformation strategies (Germany, USA and China) have two distinctive components: one component that focusses on Industrial Internet of Things which is the technologies used to transform the manufacturing industry; and the other that focusses on industry 4.0 which is how technologies are used to transform the manufacturing industry. Unfortunately, none of them clearly differentiate these two components. This has the potential to cause confusion in literature and among participant of the value chain. In this book we use the term Industrial Internet of Things (IIoT) to mean the technologies that are expected to transform industry, and the term Industry 4.0 to mean the use of IIoT to transform Industry.

Industry 4.0 depends on a combination of many elements, including distributed intelligence, network security, massive data, cloud computing, and analytics, among other things. These elements are operationalized by the "Industrial Internet of Things", a term that generally indicates a comprehensive portfolio of seamlessly integrated hardware, software and technology-based services, with the aim to enhance manufacturing productivity and improving efficiency. Typically, Industrial Internet of Things (IIoT) enables the gathering of extensive amount of data from production lines and plants, electrical and energy systems, water and sewerage systems, business information systems, as well as buildings management and automation systems, which are generally distributed. The gathered data is transmitted to analysis centers where it is transformed into information and used to make better informed decisions. Ultimately, IIoT enables Industry 4.0 to have the following benefits: improved safety, increase uptime, lower energy costs, and improved maintenance; all of which lead to manufacturing competitiveness in cyber-physical production systems supported by Smart Grid implementations; by providing the following:

[144]

- **Real-time visibility** – The use of shop floor sensors and smart machinery gives managers greater visibility (vertical integration – see Figure 8.1). This provides them with real-time monitoring that lets them know which parts of their operation is underperforming. As well as informing them where greater efficiencies can be achieved. IIoT helps businesses identify potential problems before they occur. Rather than arbitrarily deciding when equipment needs to be replaced, connected sensors can highlight problem areas so organizations can replace their industrial equipment at the optimum time.
- **Connectivity** – IIoT breaks down workplace silos no matter how large the organization is (horizontal integration at the plant and organization level – Figure 9.1). It connects machines to one another, enabling managers to get a holistic view of their industrial output. In addition, IIoT allows horizontal integration of organizations (see Figure 9.9), the so called value chain participants, enabling managers and decision makers (both humans and machines) to have value-chain wide visibility and optimization.
- **Data** – The IIoT generates huge quantities of data. When analyzed (through data analytics), the generated information can help companies to make practical decision in real-time, create new business models and identify additional revenue streams.
- **Automation** – IIoT and the proliferation of smart devices also leads to further automation, which enables businesses to maximize their productivity levels. In addition, industrial automation may also deliver safer working environment, with machines able to self-monitor and identify any potential risks.

## 9.2    Network Infrastructure of IIoT

There are many industrial networks protocols that were design to address various industrial communication challenges. Therefore, it is common to have multiple networks in a manufacturing plant. Figure 9.1 shows an industrial network of a plant that has process (e.g food processing) and electrical (including substations) automation systems. Such a plant could also have a building and environmental management automation system, which would be integrated with the process and electrical systems by the network. At the unit level, networks need to be isolated to increase communication speed and determinism, and to avoid mistakes that may result into sending wrong commands to process components. On the other hand, at group level, (process and substation networks in Figure 9.1) networks may be integrated through interfaces or routers to enable horizontal integration within the plant. Otherwise, horizontal integration is achieved through the plant network at the SCADA level.

At the plant level of the industrial network architecture, devices share large amounts of information, which necessitates high capacity networks. Therefore, most plant level networks use Ethernet based protocols such as PROFINET, Ethernet/IP and Modbus TCP, since Ethernet is known to carry large amount of data. Moreover, using Ethernet based protocols at the plant level makes it easy to integrate the plant, business, electricity and energy monitoring, and building automation networks. This is a feature of plant networks or Ethernet to be more specific, which is highly desirable in the implementation of Industry 4.0 paradigm. Like the general manufacturing, industry, the electricity industry is being affected by the fourth industrial revolution. Consequently, the grid has become smart leading to the need for information sharing among various energy centers. Standards such as IEC61850 have been developed to take advantage of the strength of Ethernet based protocols to streamline the sharing of data and information among electricity network devices and energy centers.

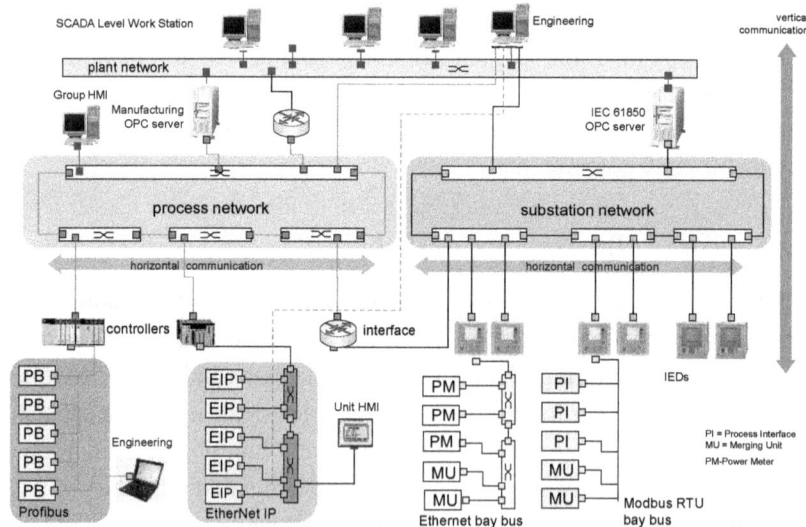

**Figure 9.1: Process and Electrical Automation Network**

Another advantage of Ethernet is its ability to support multiple communication protocols on the same wire (network). Figure 9.2 shows that manufacturing networks (PROFINET and Ethernet IP), building automation network (BACnet), electricity substation automation network (IEC61850), and general purpose network (Modbus TCP) can be supported by the same Ethernet network. Therefore data from different systems and from other value chain participants can be integrated in a single repository and processed together to generate high quality information used in decision making and in the improvement of process and plant control models.

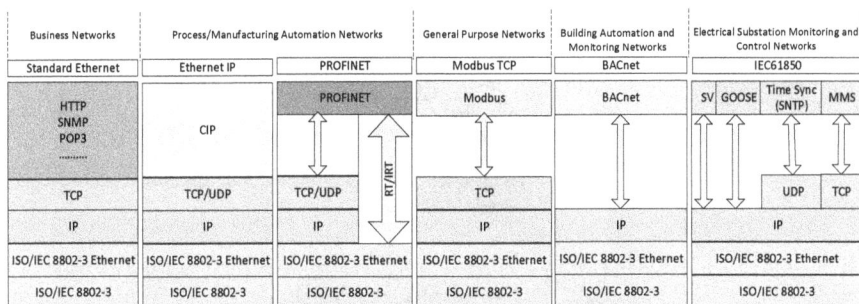

| Business Networks | Process/Manufacturing Automation Networks | | General Purpose Networks | Building Automation and Monitoring Networks | Electrical Substation Monitoring and Control Networks | | | |
|---|---|---|---|---|---|---|---|---|
| Standard Ethernet | Ethernet IP | PROFINET | Modbus TCP | BACnet | IEC61850 | | | |
| HTTP SNMP POP3 ... | CIP | PROFINET | Modbus | BACnet | SV | GOOSE | Time Sync (SNTP) | MMS |
| TCP | TCP/UDP | TCP/UDP (RT/IRT) | TCP | | | | UDP | TCP |
| IP | IP | IP | IP | IP | IP | | | |
| ISO/IEC 8802-3 Ethernet | ISO/IEC 8802-3 Ethernet | ISO/IEC 8802-3 Ethernet | ISO/IEC 8802-3 Ethernet | ISO/IEC 8802-3 Ethernet | ISO/IEC 8802-3 Ethernet | | | |
| ISO/IEC 8802-3 | ISO/IEC 8802-3 | ISO/IEC 8802-3 | ISO/IEC 8802-3 | ISO/IEC 8802-3 | ISO/IEC 8802-3 | | | |

**Figure 9.2: Using Ethernet to Support Multiple Communication Protocols**

## 9.3 Software Infrastructure of IIoT

IIoT depends on software applications to integrated data from various sources that support and affect the manufacturing industry. Such software applications include Enterprise Resource Planning (ERP) Systems, Manufacturing Execution Systems (MES), database applications, data analytics applications, and systems control logic applications. Figure 9.3 presents the architecture of the software infrastructure of IIoT. It shows that data is transported to and from the primary IIoT applications using Commercial Off-The-Shelf software based on OPC technology, or custom made applications based on web services and Dynamic Data Exchange (DDE) technologies.

An OPC servers such KEPServer can be used to collect data from control systems and send it to data and historical data access applications, as well as alarms and events applications. Moreover, KEPServer, as well as OPC clients such as DataHub support DDE with databases and data processing applications such as excel and MATLAB. To implement DDE, and web services that integrate local, and remote databases located on cloud servers, custom software applications may have to be developed. Note that KEPServer OPC server supports OPC Unified Architecture which is based on web services. Therefore it can be used to integrate applications over the Internet.

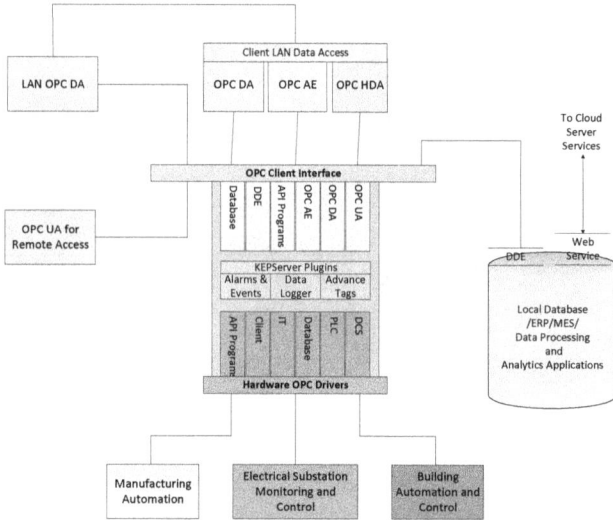

**Figure 9.3: Architecture of the Software Infrastructure of IIoT**

## 9.4    IIoT Example Laboratory

In order to train the next generation of engineers that are ready to support Industrial Internet of Things, we have developed a number of industrial networks laboratories. One of these laboratories aims at the integration of process automation systems and electrical automation systems, based on the network hierarchy shown in Figure 9.1. Figure 9.4a shows the actual network architecture of the laboratory. In the lab an Automation Direct CLICK micro PLC is configured to reading electrical parameters (voltage, current, power, and energy) from a power meter, through a Modbus RTU connection. Figure 9.5 shows the configuration of the CLICK micro PLC Modbus ladder logic instruction used to read the power meter registers. The instruction is powered through a timer to read the registers once every second. This reduces the traffic on the Modbus network. Before programming the communication, it is important to draw a logical representation of the communication among the devices on the network; and Figure 9.4b is a hand

[148]

sketch of a logical representation of the communication among devices on the network shown in Figure 9.4a.

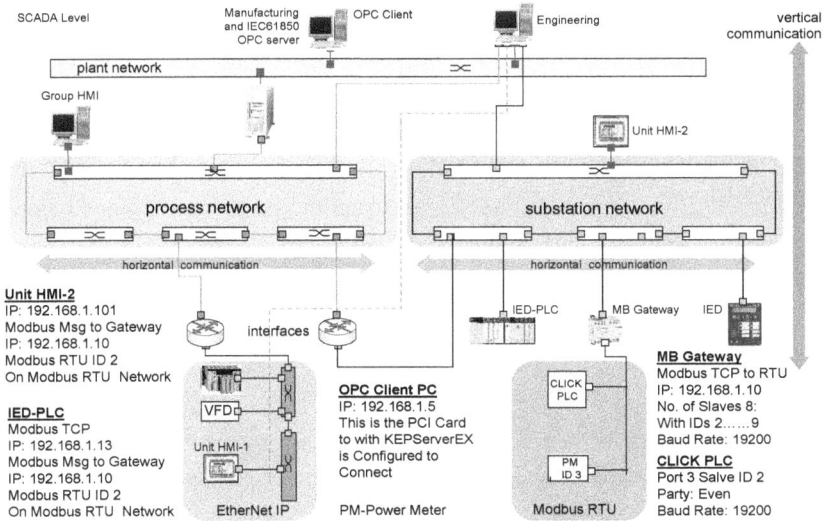

Figure9.4a Architecture of a Process and Electrical Automation System Industrial Network Laboratory

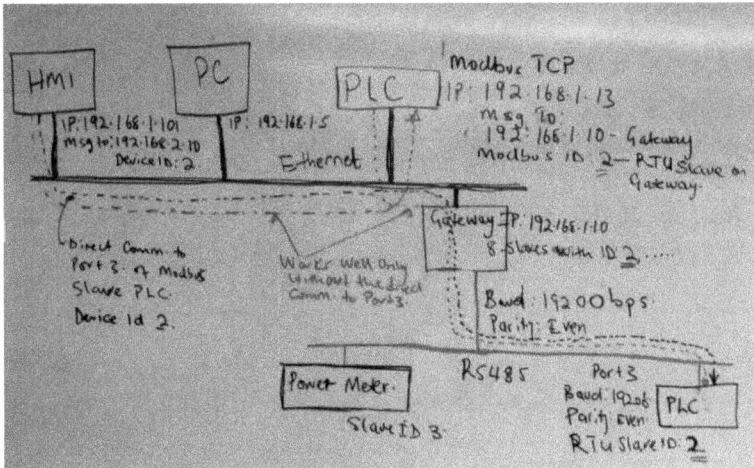

Figure 9.4b: Hand sketch of a Depiction of the Messages in Figure 9.4a

**Figure 9.4: Modbus TCP and Modbus Serial Networks**

The HMI connected to the substation network in Figure 9.4 can read the power parameter from the CLICK micro PLC through the Schneider TSXETG100

Modbus RTU to Modbus TCP gateway. In this case, each register is read separately by a HMI instruction configured as shown in Figure 9.6.

Since HMIs do not support logic instruction, the registers are read periodically, causing a great amount of traffic on the Modbus RTU network. This causes the HMI to flag a message timeout error from time to time. This issue is addressed by using a Productivity 3000 PLC as an Integrated Electronic Device (IED) to read the CLICK registers through the Modbus gateway, using a Modbus TCP read instruction. The configuration of this instruction is presented in Figure 9.7, showing the IP address of the gateway, as well as the Modbus RTU address of the CLICK micro PLC.

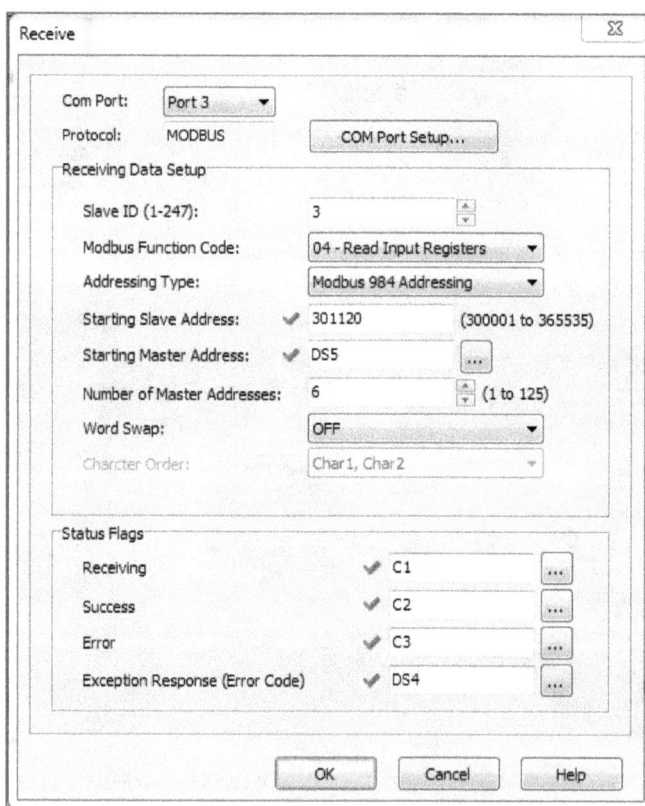

**Figure 9.5: Configuration of CLICK micro PLC Modbus-Read Ladder Logic Instruction**

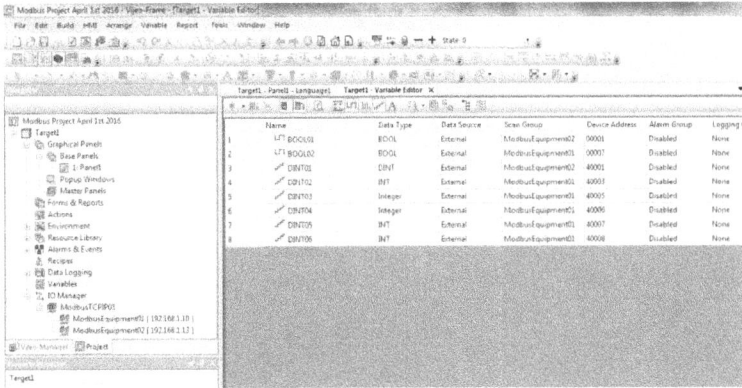

Figure 9.6a: HMI Variables

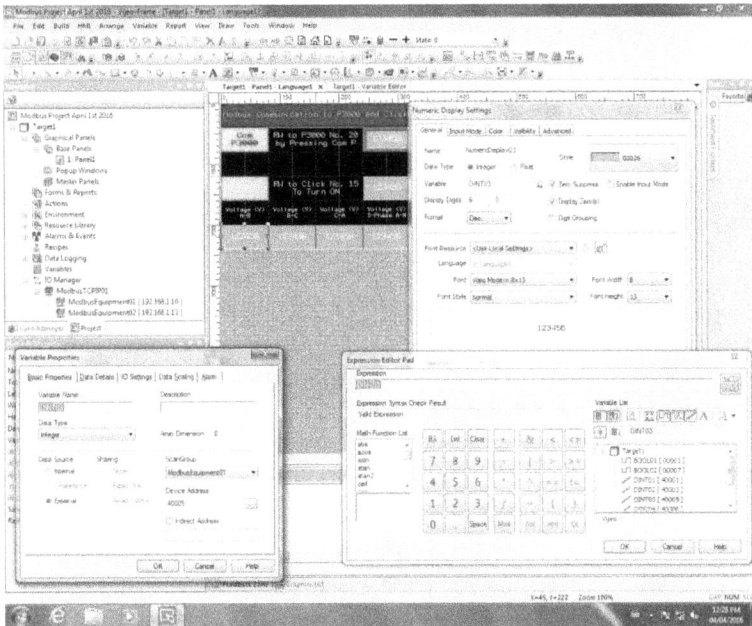

Figure 9.6b: HMI Variable Configuration

**Figure 9.6: HMI Implementation**

Note that the Productivity 3000 PLC (IED) can be used to read the power meter registers directly through the gateway using a Modbus TCP gateway, or it can be connected to the Modbus RTU through one of its RS485 ports.

[151]

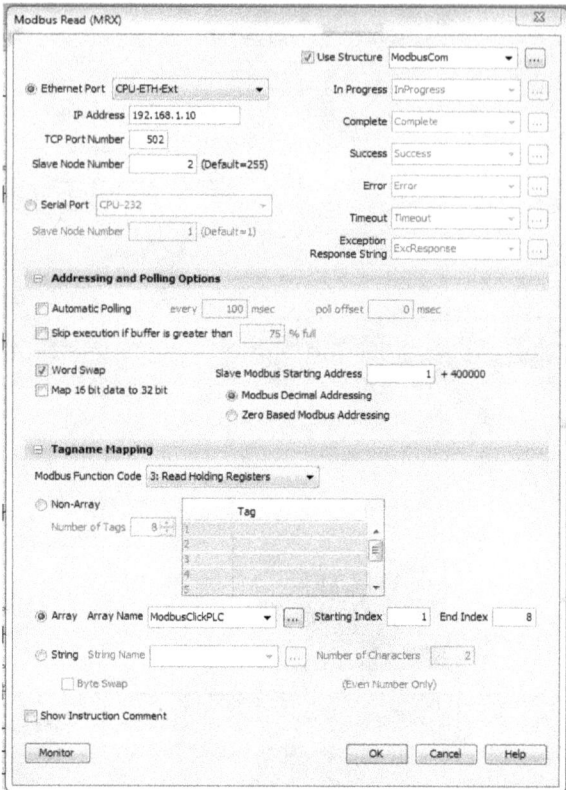

**Figure 9.7: Productivity 3000 Modbus Read Instruction**

The OPC based SCADA client in Figure 9.4a accesses the SEL 751A relay electrical data through an IEC61850 OPC server. In addition, the client reads the Schneider PM800 power meter data from the Productivity 3000 PLC through a Modbus TCP OPC server that runs on the same PC with the IEC61850 server. The Modbus TCP OPC server can read the electrical parameters directly from the power meter using the channel configuration shown in Figure 9.8. However, the link between the server and the meter has a Modbus RTU segment which does not support connection orient communication. This results into intermittent loss of connection between the server and the meter, which in turn is intermittently flagged by the OPC client.

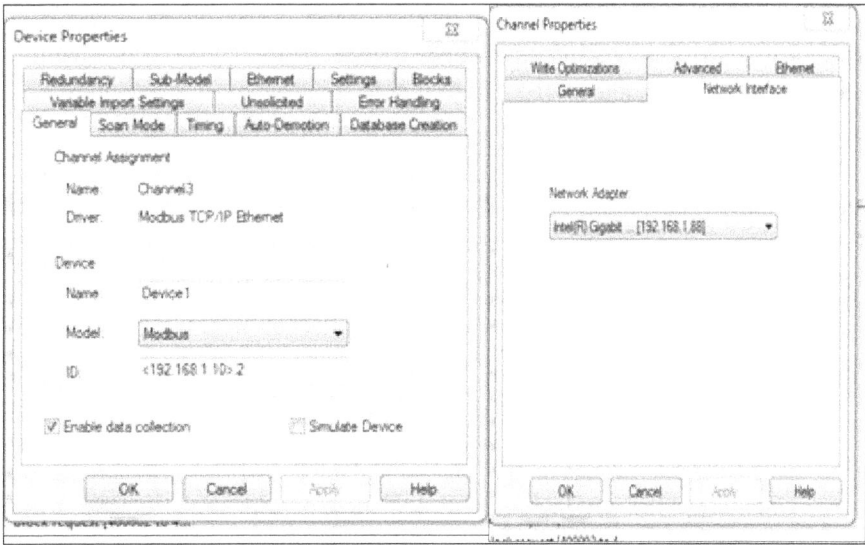

**Figure 9.8: Modbus OPC Channel Configuration**

On the process side, the laboratory uses a Productivity 3000 PLC, Eaton VFD, and Eaton **ELC-CAENET Remote** IO module (see Figure 9.4a). This equipment communicate using Ethernet IP, and it is the focus of the laboratories presented in Appendices A and B. The Productivity 3000 PLC communicates with the remote IO using Ethernet IP implicit messaging, and with the VFD using Ethernet IP explicit messaging. The details of the network configuration is presented step by step in Appendices A and B. We have also developed an online version of the lab that uses productivity 3000 PLC and Powerflex VFD. This lab is available on request.

The laboratory demonstrate the availability of technologies that allow the integration of industrial automation devices from various manufactures to create an Internet of Things. This Internet of Things does not mean industrial transformation from the business point of view, but it can be used together with appropriate business models to create industrial transformation.

## 9.5 Industry 4.0

The deployment of technologies such as industrial networks or Internet of Things technologies, cloud computing and big data software applications, smart manufacturing technologies such as CNC machines and 3D printing, does not in

[153]

itself amount to Industry 4.0. Because Industry 4.0 is not the technologies, but the use of the technologies (i.e. integrated industrial automation, remote and intelligent system monitoring, data analytics and autonomous decision-making; all of which are supported by real-time or almost real-time communication at all levels) to optimize production and logistics. There is a lot of definitions as well as technology and policy proposals of Industry 4.0 in literature. But at the operational level, Industry 4.0 is not about the technologies, but about a systems approach to production and logistics optimization. Industry 4.0 technologies simply provides the following:

- Expansion of the system to include the entire value chain (see horizontal integration of the value chain in Figure 9.9), as opposed to being just a single plant or organization, through industrial networks, cloud computing, as well as global and local databases.
- Capability for production and logistics optimization through system monitoring, data analytics and autonomous decision-making.
- Capability for efficient manufacturing through increased automation and smart manufacturing technologies such as CNC machines and 3D printing.

Figure 9.9 shows that Industry 4.0 technologies enable organizations to automatically collect, store and analyze relevant information from all their value chain participants. This process depends heavily on Industry 4.0 technologies and is general for organizations that intend to implement Industry 4.0 manufacturing paradigm. On the other hand, the optimization of production and logistics is specific to organizations. It depends on the organizations' Industry 4.0 objectives and implementation strategies. Therefore, the applications that are used by organizations to achieve their Industry 4.0 objectives have to be customized to the organizations and to their objectives. The customization process must follow a systems approach to capture all the necessary information.

[154]

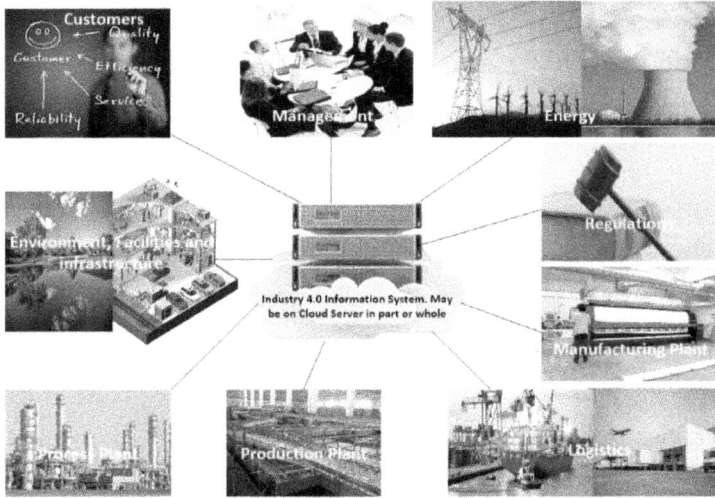

Figure 9.9: Horizontal Integration of Value Chain Participants

## 9.5.1 Laboratory and Example of the Implementation of Industry 4.0 Technologies

This example uses the laboratory equipment whose architecture is shown in Figure 9.10. The equipment represents a fictitious heat exchanger whose temperature is controlled by a PID that is run in automatic or manual mode. The PID controller is deployed on a Micrologix 1400 PLC. The electrical system of the heat exchanger is monitored by a SEL751A relay, and the building lighting system is controlled by a PXC building automation controller. All controllers are connected to the same remotely located OPC server through a VPN connection, despite the fact that Micrologix 1400 PLC communicates using Ethernet IP, SEL751A uses IEC61850, and PXC controllers uses BACnet IP. The laboratory work based on this equipment focuses on training students the integration of data from different value chain participants using OPC technology, and using integrated data to control manufacturing processes. In other words the equipment is used to teach horizontal integration of industrial automation systems (Figure 9.1).

The process/manufacturing, electrical, and building automation data of the system in Figure 9.10 is accessed from the OPC server by a client that runs on the same PC. The PC is located at the SCADA level of the automation hierarchy. Furthermore, the equipment is used in a course project to support students' work in the development of web services and DDE applications, and to develop HMI

[155]

such as the one shown in Figure 9.11. This HMI enables operator to tune the PID, put the PID in automatic or manual mode, set heat exchanger temperature, turn lights on and off, and start/stop the temperature control process. It also enables the machine operator to monitor the status of the electrical power supply to the process. These functions of the HMI are supported by the vertical integration of the automation system. The horizontal integration of the process, electrical, and building automation systems enable the operator to set a power consumption above which the lights cannot go on. Many such energy saving scenarios can be accomplished due to the vertical and horizontal integration of automation systems. Usually these scenarios are implemented at the Manufacturing Execution System (MES) level of automation systems. Therefore it is necessary for the next generation of engineers to know how to move data between the SCADA and the MES levels.

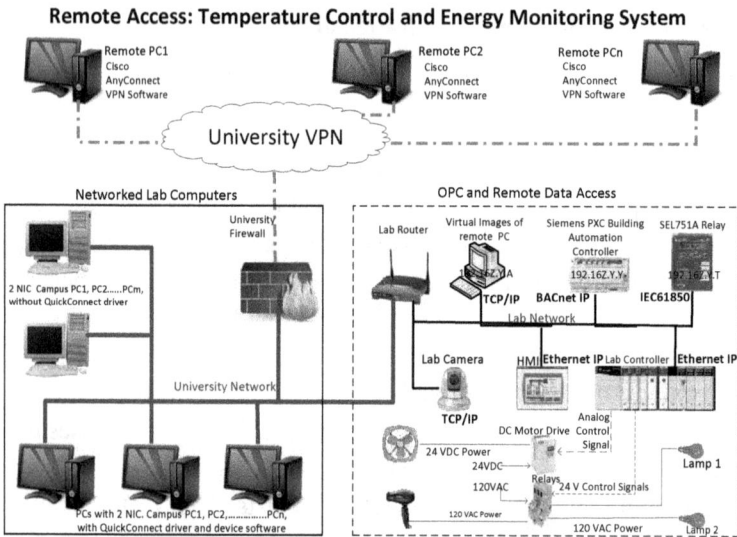

**Figure 9.10: IIoT Lab Architecture**

Note that because of horizontal integration of automation systems, we can have a single group HMI (Figure 9.11) that is used to operate the temperature control system and the building lighting system, In addition, the HMI is used to monitor the electrical power supply. If such an HMI is deployed in a factory settings, different operators would be given different access rights to different part of the HMI.

[156]

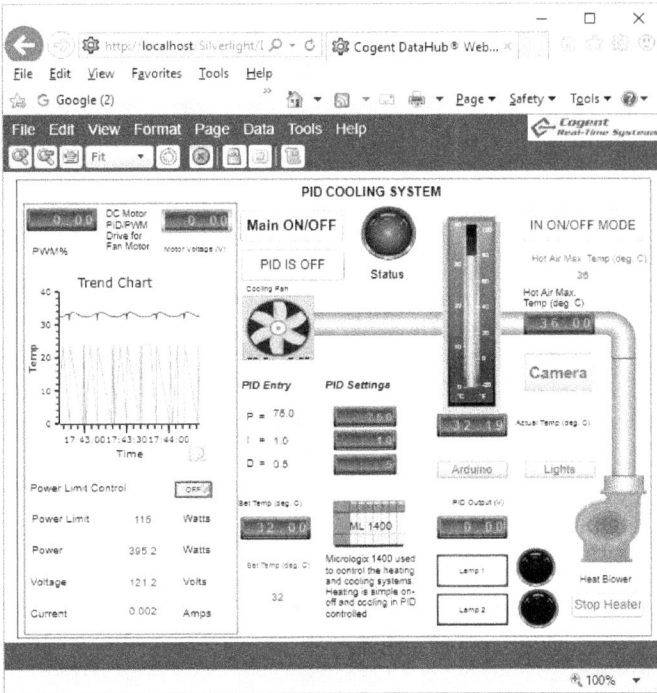

**Figure 9.11: Group HMI for Temperature Control and Electrical System Monitoring**

## 9.5.2 Example of the Implementation of Industry 4.0 Paradigm

We use the Industry 4.0 example presented in Section 9.5.1 to show how a systems approach is used to implement Industry 4.0 complaint system for monitoring and controlling industrial plants. At this moment we know that our system can send process, electrical and building automation data to a single database or application through OPC DDE. Therefore, our next step is model the Industry 4.0 implementation problem as follows:

- Establish objectives that should be met by the implementation of Industry 4.0 paradigm.
- Determine the data points from the automation systems that are used to generated actionable information
- Determine the data from the value chain participants that is used in the analysis and decision-making that leads to meeting the set Industry 4.0 objectives.

[157]

- Develop an application that automate the Industry 4.0 decision-making in full or in plant.

In this example we model the activities of example problem using the Integrated DEFinition Methods (IDEF). IDEF is the common name referring to classes of enterprise modeling languages. It is used to model activities necessary to support system analysis, design, improvement or integration.

A component of IDEF called IDEF0 originated in the U.S. Air Force under the Integrated Computer Aided Manufacturing (ICAM) program in the 1970's; from a well-established graphical language, the Structured Analysis and Design Technique (SADT). IDEF0 is a function modeling method for analyzing and communicating the functional perspective of a system. Its models do not only help to organize the analysis of a system, but also promote good communication between the analyst and the customer. IDEFØ is useful in establishing the scope of an analysis, especially for a functional analysis. As an analysis tool, IDEFØ assists the modeler to identify the system functions, the data and methods needed to perform the functions, and the systems reliability.

In our example we start by establishing our Industry 4.0 objective as reducing the cost of energy used to run the temperature control system and the lighting system of a fictitious building where our heat exchanger is located. Using IDEF0 modelling techniques, we place our objective in the middle of a box A0 as shown in Figure 9.12. Through a brain storming exercise we determine the output (saved energy cost) of the system as well as a set of inputs to the system. The input to our system has the following characteristics that are important during the analysis stage:

- Lights control and temperature setpoint are inputs made by operators of the system
- Measured temperature comes from the heat exchanger temperature sensors
- Time of Day (ToD) energy pricing comes from the electricity utility company. Since this data is out of our control, it could change without warming, leading to erroneous decisions. Therefore we decided to obtain it from the company's website in real time whenever the system needed it.
- Energy saving strategies comes from our knowledge of the usage of the heat exchanger and building lighting system. This helps us to determine the essential and the none-essential load.

[158]

Up to this point we do not know the data set to collect from the automation system as well as the analysis we need to carry out on the data so as to generate energy cost saving information. Therefore, we open up A0 and identify the activities of the system that coverts the inputs to the identified output of saved energy cost. This results into Figure 9.13, which shows the data such as power, current, and energy saving baselines, power factor, and temperature that we need to generate information that we can use to save energy.

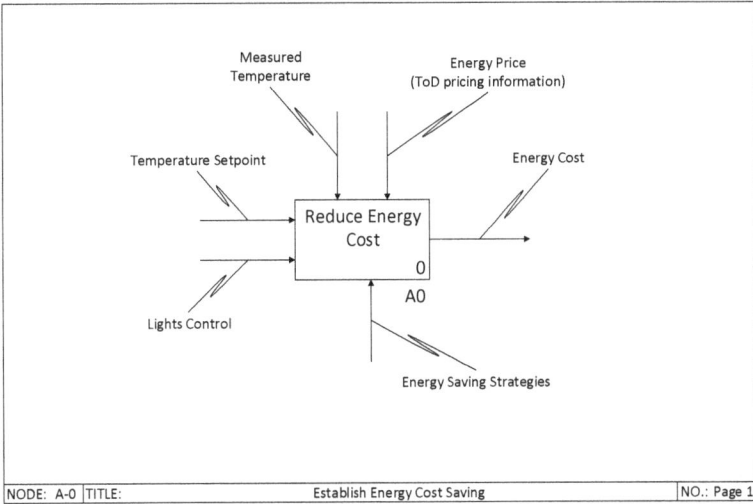

Figure 9.12: IDEF0 Model A0 for Reducing Energy Cost of Running a Heat Exchanger and its Building

This data is passed on from the automation system to the excel document shown in Figure 9.14 using OPC DDE. We also use the excel document to automatically read the ToD electricity pricing from the utility's website and update the reading at the top of every hour. Once all the data is in the excel document we use Visual Basic macro programming to generate tables and graphs that help decision maker to develop good energy saving strategies.

Once the strategies are agreed up on, they are implemented in the control applications. The baseline data is entered into the system at the SCADA level and distributed to all system components.

[159]

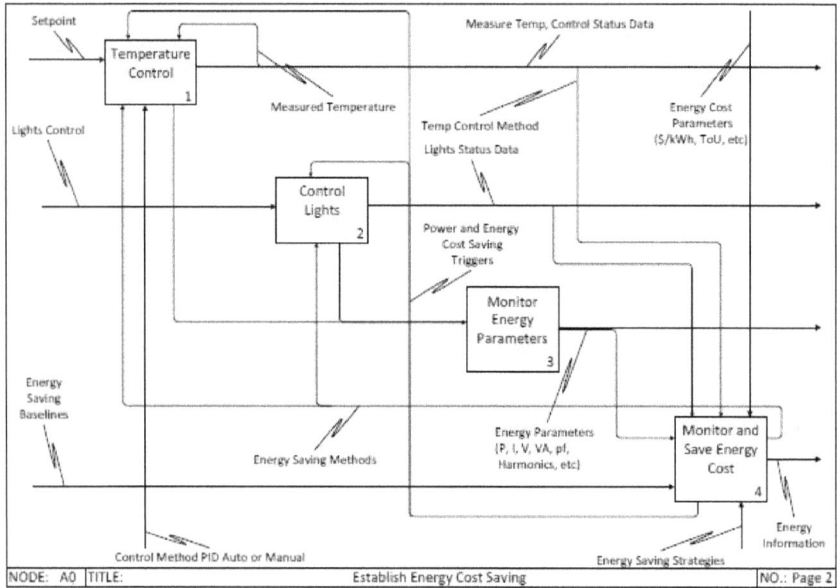

Figure 9.13: Analysis of a Model for Reducing Energy Cost of Running a Heat Exchanger

In our example, we use the strategy of turning off the none-essential load (the light) during peak hours, and whenever the power consumption goes above a preset energy saving baseline. We also tested the energy consumption of the temperature control system under automatic and manual PID control. Figure 9.15 shows that the system consumes far less energy when in automatic mode. In real life situation, this would result into another strategy of having a standard operating procedure that avoids tuning the PID during peak hours.

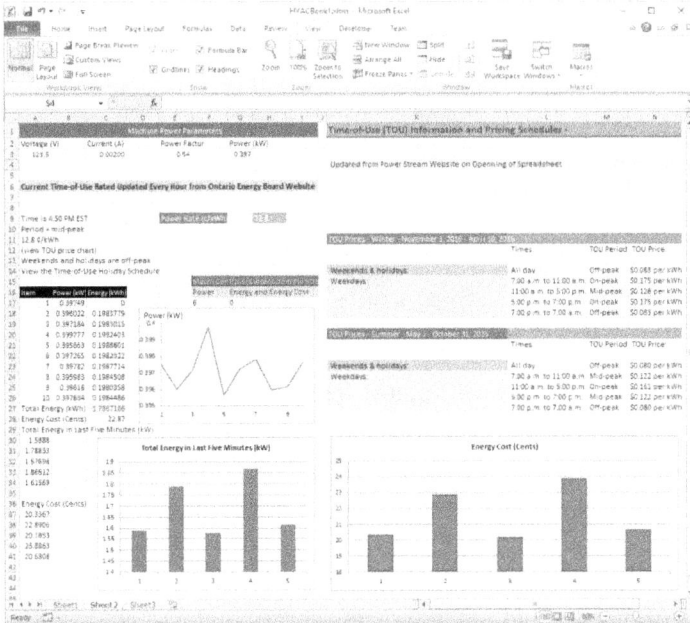

Figure 9.14: Process, Electrical, Building Automation Systems and Energy Cost Information

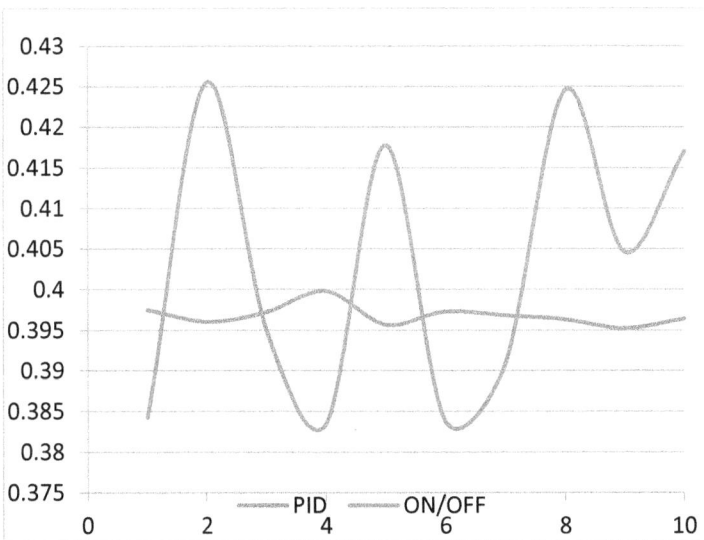

Figure 9.15: Energy Consumption of a PID Temperature Control System

[161]

This example shows that the focus of Industry 4.0 is not the technologies, but the use of the technologies to optimize processes and logistics. In addition the example shows that it is necessary to use a systems approach to integrate all the information from various value chain of organizations.

### 9.5.3 Other Examples of the Implementation of Industry 4.0 Paradigm

There is a lot of hype about Industry 4.0 in literature, focusing on complex solutions that make it financially impossible for Small and Medium scale Enterprises (SME) to participant and benefit from the paradigm. For example costly proprietary information sharing ERP applications may be out of reach for most SME, and may be very difficult to integrate with information sources of the value chain participants. Generally, I believe that good Industry 4.0 implementations are small, focused on specific objectives, inexpressive, and uses standard and proven technologies as much as possible. Therefore, like in the example presented in Section 9.5.2, the two example described in this section also focus on using standard OPC technology and Microsoft Excel application to deliver the functions of Industry 4.0.

*Example 1: Information Sharing among Companies*

In this example, think of a distributer company that packs and distributes herbicides to retailers. It is common for such companies to be unable to share information with its customer on EPR because of the following reasons:

- Lack of technology to integrate the ERP of the company and the ERPs of its customers.
- Lack of trust among participants about what others may access from their ERPs.
- Many participants may not have ERPs at all.

The business system in this example shares Excel data without sharing the Excel workbooks, because sharing the workbooks leads to many issues, including exposure of user data, application (macros) programing, and the possibility of creating multiple versions of the shared document. All these issues are avoided by using text files, ODBC database connections, or XML files to share the data. The system whose architecture is shown in Figure 9.16, is implemented as follows:

- Create an Excel document with a macros application that reads from, and writes to a text files.

[162]

- Define a local file location for the document created in step i. For example C:\Production_On_Order_Scheduling\Input_Files
- Upload the text files to cloud server and invite a retailer to access one of the text files.
- The retailers can create their own Excel documents with macros applications that automatically update the text fill on the cloud server, or they may choose to always log on the cloud server to make orders through the text file assigned to them.

Once the business system has been implemented, it works as follows to deliver the functionalities of Industry 4.0 (Figure 9.16):

- A retailer makes an order by logging in the cloud server and entering the order information into the shared text file. This information may include, product name, packaging unit (e.g. bottle, bag, or box, with associated weight or volume), and preferred delivery date. If the retailer has an Excel document that updates the workbook on the cloud server, then all they need to do to place their order is to enter it into their own document.
- The order information is automatically read by the distributor's master Excel document and organized into production data using its macros program.
- Using the product name, the master Excel document retrieve the product's recipes from the recipes database. If this was not a controlled product, the retailer could enter their own recipe as part of the order.
- Assuming the retailer ordered for a bottled product, its recipe, weight per packaging unit, and time to complete the production are automatically loaded into the bottling machine. The machine responds by automatically setting the bottling weight, dribble, and drop weights, as well as the machine speed and bottle sensor positions.
- All that remains is to install the mechanical attachments to the machine and start the production.
- During production, the bottling machine may send the following information to the master Excel documents: production start and expected completion date and time, production speed, as well as unexpected delay. Depending on the implementation, the master Excel document may send some of this information to the cloud document, hence making it available to the retailer, and helping them in their planning.

Figure 9.17 shows the implementation of the information sharing system. Orders are entered by retailers through the Excel document shown in Figure 9.17(a). When the "Send_to_Cloud" button is clicked, the order information is sent to the text file located a cloud server. On the other hand, when the "update" button in the production manager's Excel file is clicked, the information written in the text file is read into the Excel worksheet, completing the information transfer process.

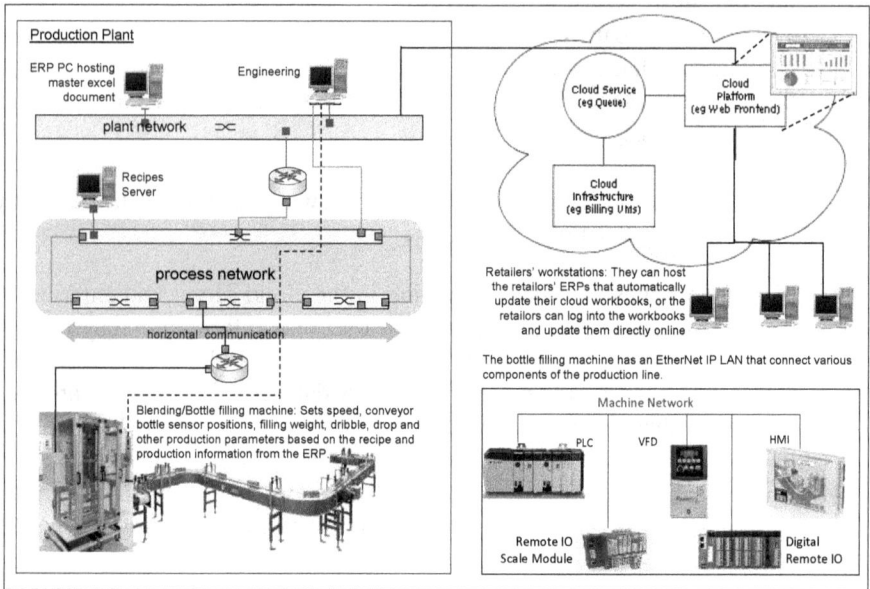

Production Plant
ERP PC hosting master excel document
Engineering
plant network
Recipes Server
process network
horizontal communication
Blending/Bottle filling machine: Sets speed, conveyor bottle sensor positions, filling weight, dribble, drop and other production parameters based on the recipe and production information from the ERP.

Cloud Service (eg Queue)
Cloud Platform (eg Web Frontend)
Cloud Infrastructure (eg Billing VMs)

Retailers' workstations: They can host the retailors' ERPs that automatically update their cloud workbooks, or the retailors can log into the workbooks and update them directly online

The bottle filling machine has an EtherNet IP LAN that connect various components of the production line.

Machine Network
PLC    VFD    HMI
Remote IO Scale Module    Digital Remote IO

**Figure 9.16: Architecture of the file sharing system**

This a simple and very effective business automation system that is based on standard technologies. Low cost combination of old technology (text files) and new technology (synchronizing desktop data over the internet). Note, that there are some data security issues that may arise from the setup of this example, but since this is not the focus of this book, it is not discussed any further. But I would like to note that information systems need to be balanced for cost, simplicity, security and performance. The unstable equilibrium that is set by complicated ERP implementations is expensive and unusable by SME.

*Example 2: Smart Automated Sprinkler*

In Canada, municipal water usage doubles in the summer months, because this is when Canadians are outdoors watering lawns and gardens, filling swimming pools and washing cars. Summer peak demand places stress on municipal water systems and increases costs for tax payers and water users; because the capacity to deliver the required water levels for the summer months must be installed even though it is used for about a quarter of the year. Much of the summer peak demand is attributed to lawn and garden watering. Therefore, as water supplies diminish during periods of low rainfall, some municipalities may declare restrictions on lawn and garden watering.

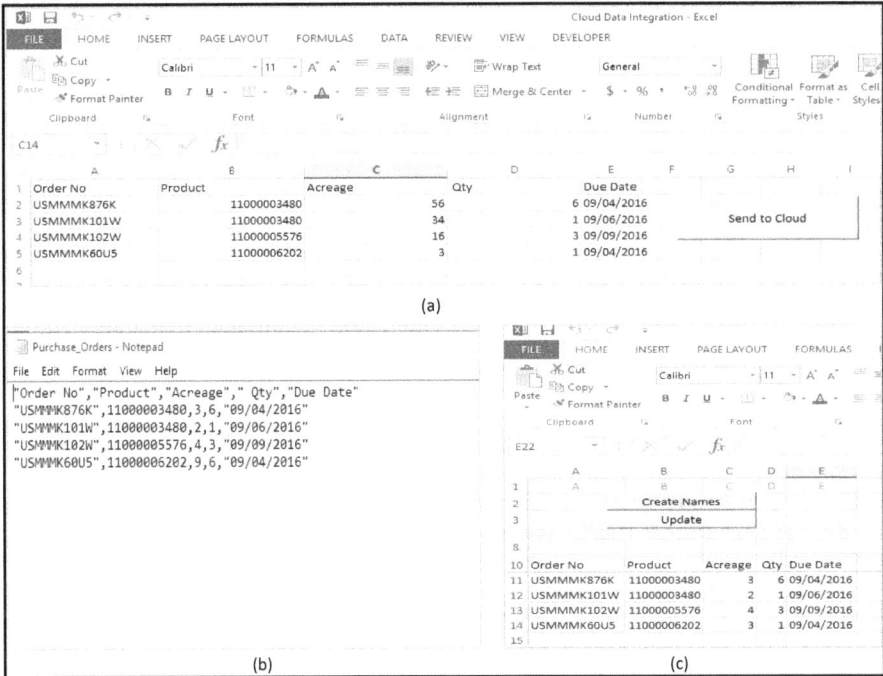

Figure 9.17: Information Sharing Implementation

The Canada Mortgage and Housing Corporation recommends that before watering, always take into account the amount of water Mother Nature has supplied to your lawn or garden in the preceding week. But in this example, our system does not only take into account the water that Mother Nature has already supplied to the lawn, but also the water she will soon supply. The smart automated sprinkler shown in Figure 9.18 is implemented as follows:

- An Automation Direct CLICK micro PLC is used to turn on the sprinkler for $T$ minutes
- An Excel file reads the weather information for the next 24 hours from a weather website (Figure 9.18). The file shown in Figure 9.19 run macros that remove units from the read data and replaces dashes with zeros. Moreover, the total amount (mm/hour) of rain expected in the 24 hours,

and the maximum chance of raining are determined and passed over to OPC DDE. This information together with the moisture level of the lawn read and sent to the CLICK micro PLC by two moisture sensors through a Bluetooth connection is sent to a Matlab application.

- The Matlab application uses fuzzy logic to integrate the data from the OPC DDE, and calculate the time T required to sufficiently water the lawn. $T$ is sent to the PLC through OPC DDE.
- At watering time, the CLICK micro PLC opens the valve for $T$ minutes, which depends on the amount of moisture already in the ground, the expected amount of moisture in the next 24 hours, and the chance that it will rain in the next 24 hour.

The membership functions of the fuzzy logic controller for amount of rain, chance of raining, soil moisture level, and sprinkle amount are shown in Figure 9.19. These functions are used to generate fuzzy values of the input data that is integrated using the following set of rules:

1. IF moisture level is HIGH, THEN watering in LOW
2. IF moisture level is MODERATE, and chance is HIGH, THEN watering in LOW
3. IF amount is HIGH, THEN watering is LOW
4. IF moisture level is LOW, and amount is HIGH, THEN watering is MEDIUM
5. IF moisture level LOW, amount is MODERATE, and chance is LOW, THEN watering is HIGH
6. IF moisture level is LOW, and amount is LOW, THEN watering is HIGH
7. IF moisture level is LOW, and chance is LOW, THEN watering is HIGH

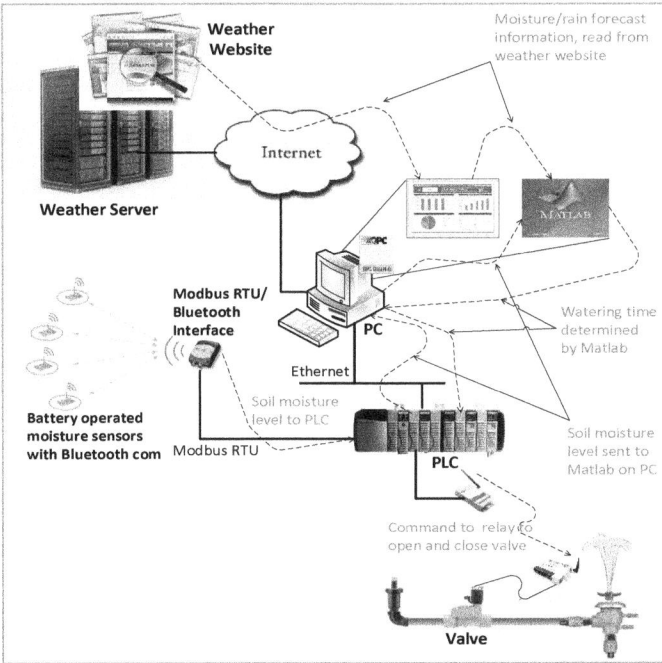

**Figure 9.18: Components of Intelligent Sprinkler System**

**Figure 9.19: Excel File with Macros Programming to Automatically Read the Weather Website**

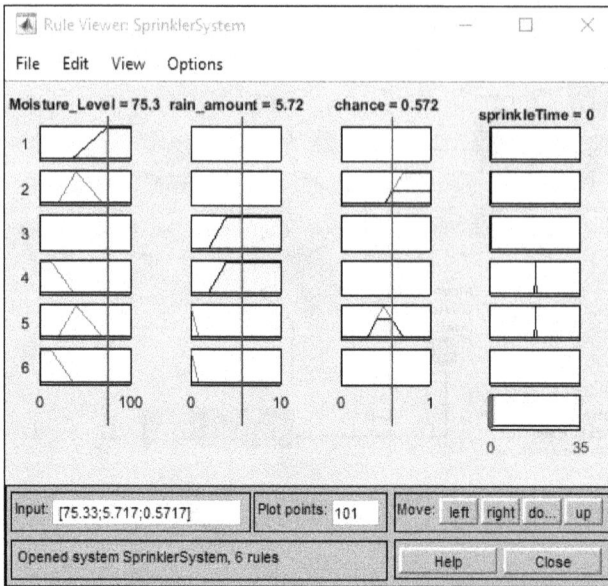

**Figure 9.20: Membership Functions of Intelligent Sprinkler Input Data**

The rules are entered into Matlab through an interface shown in Figure 9.21. A diagrammatic representation of the rules in Figure 9.22 allows the fuzzy logic controller designer to fine tune the rules to meet the expert and correct operation of the system.

For example Figure 9.22a shows that a bad rule had been made that would turn on the sprinkler when there is a lot of moisture in the soil. On the other hand Figure 9.22b shows that when there is little moisture in the soil and it is not expected to rain in the next 24 hours, the lawn is watered for a long time; and if there is a lot of moisture in the soil, the sprinkler is not turned on.

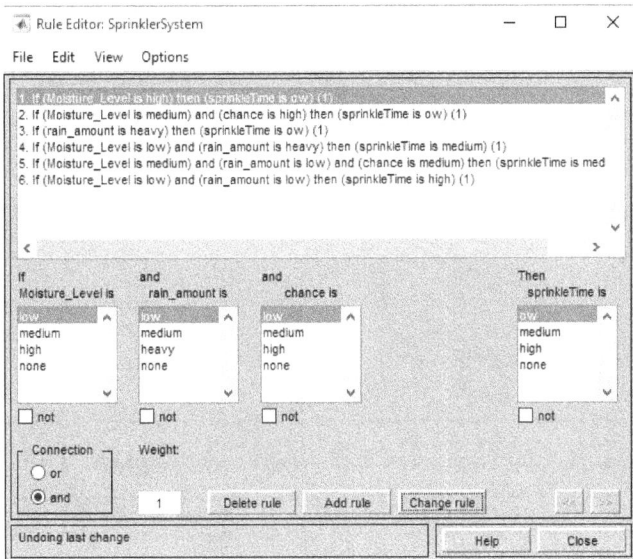

Figure 9.21: Rule Entry Interface

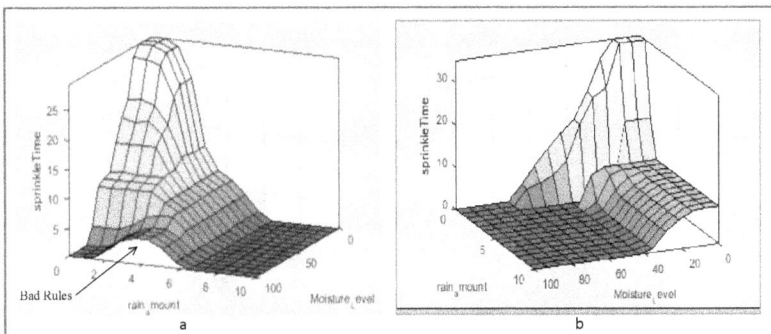

Figure 9.22: Diagrammatic Representation of the Intelligent Sprinkler Fuzzy Logic Controller

Before deploying the system, it is simulated using the Simulink model in Figure 9.23. In the simulation rain amount, chance, and soil moisture level are assumed to be random. This generates the results in Figure 9.24, which show that the watering time of the intelligent sprinklers was above 20 minutes less than 20% of the watering sessions; as opposed to water for 20 to 30 minutes (100%) every session.

[169]

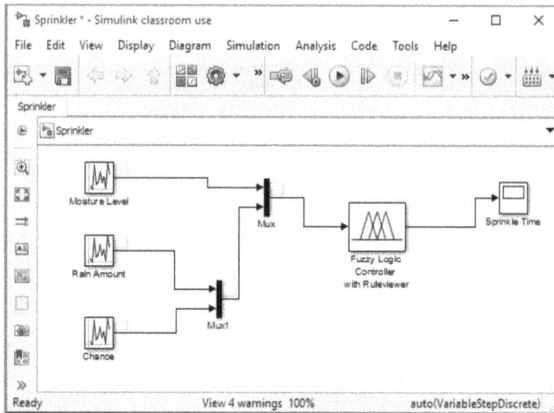

**Figure 9.23: Simulation of the Intelligent Sprinkler System**

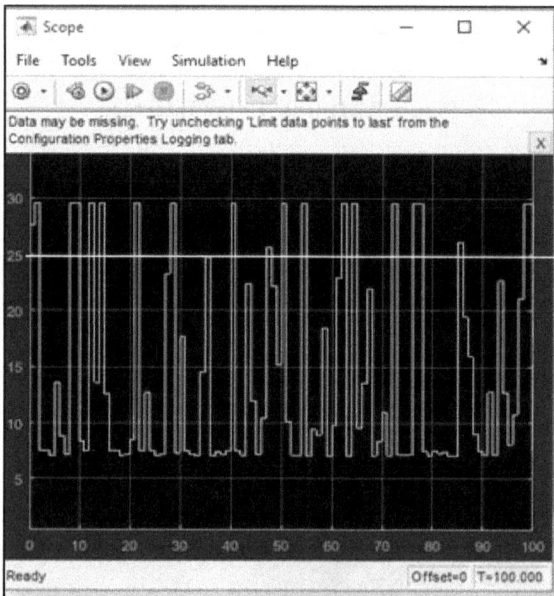

**Figure 9.24: Results of the Simulation of the Intelligent Sprinkler System**

## 9.6     Summary

This final chapter focuses on the relationship between Industry 4.0 and Industrial Internet of Things (IIoT). We make it clear that these two concepts are not the same, but one IIoT servers that other. We present a comprehensive example of the configuration on industry IIoT as well as three simple and low cost Industry 4.0

solutions. This is not to say that complex solutions are not required in Industry 4.0, but to highlight the key issue of Industry 4.0, which is the use of IIoT to increase, simplify, and optimize manufacturing. This is something that is glazed over by most in literature as they rush to present complex, costly, and high level solutions that are virtually impractical for most SMEs. We believe that this approach will help industry to benefit from the low-lying fruits of Industry 4.0 that are based on low cots technologies, as well as hardware and software systems that are already widely deployed.

## 9.7    References

[1]    Integrated DEFinition Methods (IDEF) IDEF, Family of Methods-A Structured Approach to Enterprise Modeling & Analysis, Accessed from: http://www.idef.com/

[2]    Tony Rice, Production Planning and Scheduling by Spreadsheet, Production Scheduling, Accessed from: http://production-scheduling.com/

## 9.10 Discussion Questions

**Question 1**

In the factories of the future, work pieces will contain information in the form of production parameters. As the parts approach the correct station, they tell it which part and variant they are, and request to be processed with the appropriate method. In line with this production paradigm, design an Industry 4.0 enabled product blending system for which the recipe is stored in the product containers which communicate with fill stations through RFID tags. The system should have inventory management as well as information sharing capabilities.

**Question 2**

Describe how a piece of manufacturing equipment can use Industry 4.0 technologies to provide smart service and maintenance functionalities.

**Question 3**

Design a manufacturing energy monitoring and saving scheme that involves benchmarking of the energy consumption of the plant, which that of other similar plants.

# *Appendices*

## Appendix A

### Lab A: Configuration of Ethernet IP Devices

#### A.1 Objective

The purpose of this laboratory is to learn various methods of configuring communication of Ethernet IP devices.

#### A.2 Material

The following materials are required for this laboratory

i)     Eaton Ethernet IP – Smartwire Unit

ii)    Personal Computer ("the PC") with Anybus Ipconfig for variable frequency drive, Eaton EICSoft for the remote I/O device

#### A.3 Procedure – Part 1: Assigning IP Addresses Using Configuration software and Dip Switches

#### A.3.1 Assign IP address using configuration software, LAN (local area network)

i)     Connect the Ethernet/IP devices or modules to Eaton Ethernet switch (Power Xpert Ethernet Switch) to any of the six ports available.

ii)    Using an Ethernet cable, connect the Ethernet switch to your computer. Your network should be similar to Figure A.1.

iii)   Plug in the power cable of the EATON board and turn on your computer.

iv)    You need to change the IP address of the local network. To view Network Connection click start →in the search bar→ type ncpa.cpl

v)     Once the Network Connection window opens, right click on the LAN network "Local Area Connection2" → and click properties. Note that "Local Area Connection2" should be replaced by the Ethernet connection used to connect to the hardware.

vi)    Select (TCP/IPv4) and click properties :

**Note.** Record the original settings as you need to change them back at the end of the lab.

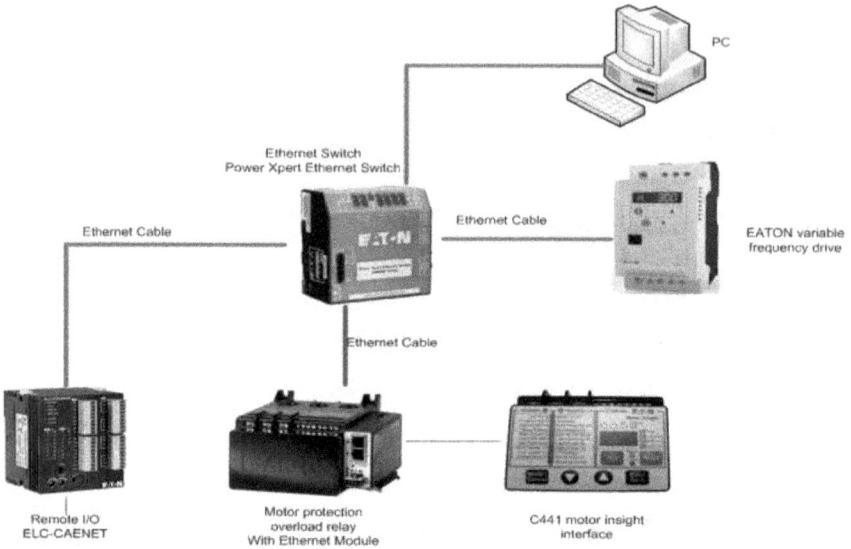

**Figure A.1: Physical Ethernet Network**

Next → select (<u>use the following IP addresses</u>) option and change the IP address to the following:

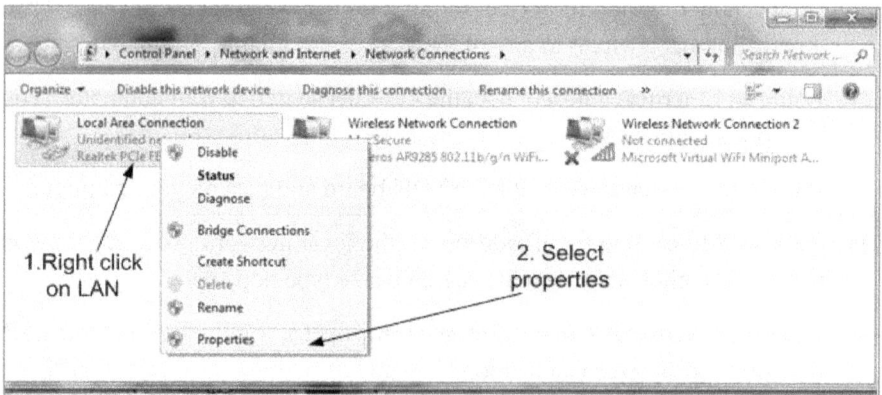

The window below opens;

**IP Address:** 192.168.1. **X**

**Subnet mask:** 255.255.255.0

**X** is your computer station number. For example if your computer station is PAT-6, then your IP address is 192.168.1.**6**.

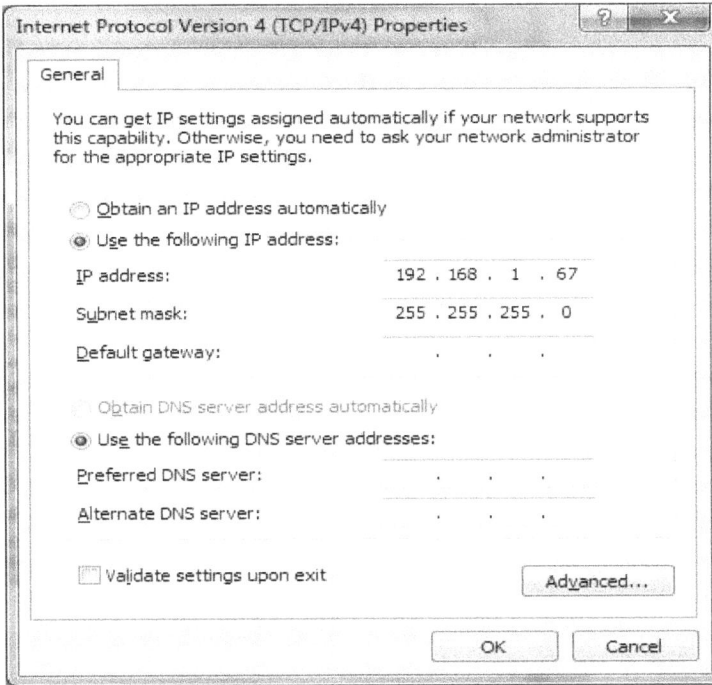

**Note.** You could select any other number as long as it is not used somewhere else in the network.

### A.3.2    Assign and IP address to DA1 [2]

i)    To assign an IP address to DA1 (variable frequency drive), Anybus IPconfig software is used. Open Anybus Ipconfig as follows:

Start → programs → HMS → Anybus IPconfig

ii)    The program usually finds the device automatically. If the device/module is not shown click on Scan. The program will show all available modules.

iii) Right-click on the line for the module and select the Configuration option from the context menu in order to assign the module an IP address.

Now set an IP address of the DA1 to:

**IP Address**: 192.168.1.1X

**Subnet mask**: 255.255.255.0

Confirm with **OK**.

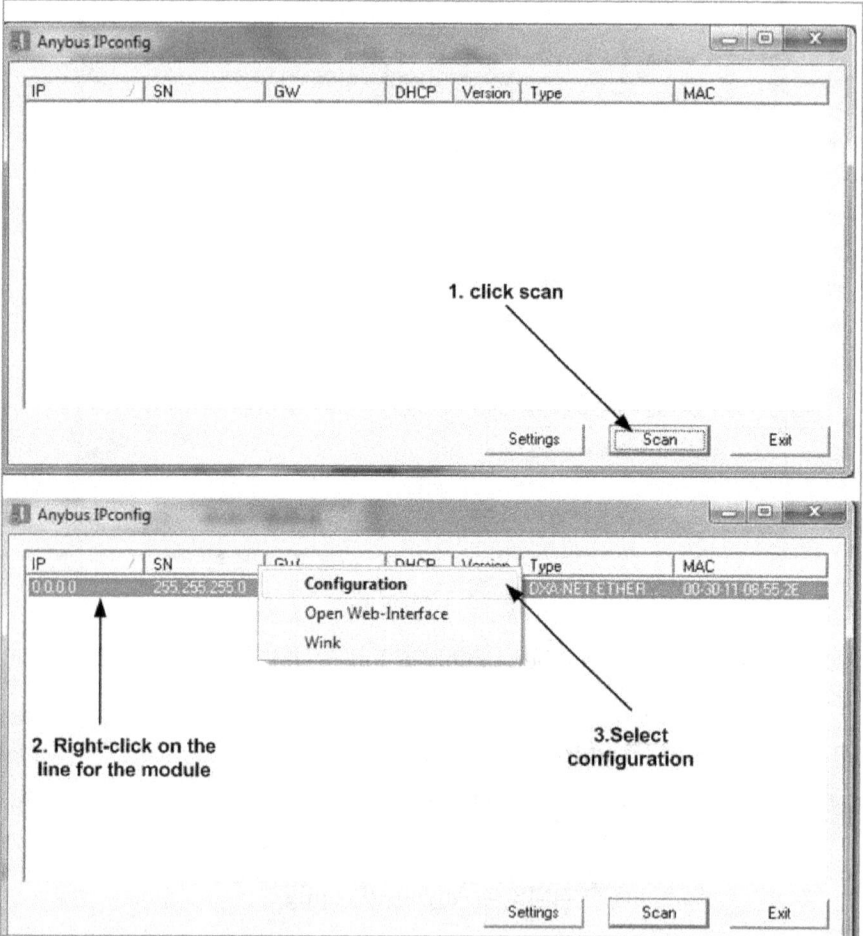

If your station is PAT 6, now the device (DA1 –variable frequency drive) has an IP address of 192.168.1.16

### A.3.3    Assign and IP address to: ELC Distributed I/O (ELC-CAENET) device [3]

i) To assign an IP address to the remote I/O, Eaton ECISoft software is used. To open the software:

Start → programs → Eaton→ Communication →ECISoft

ii) Open ECISoft and click on the "IP Search" icon in the toolbar

iii) The communication module will be displayed when found.

iv) Double-click on the module to be configured to enter the setup page.

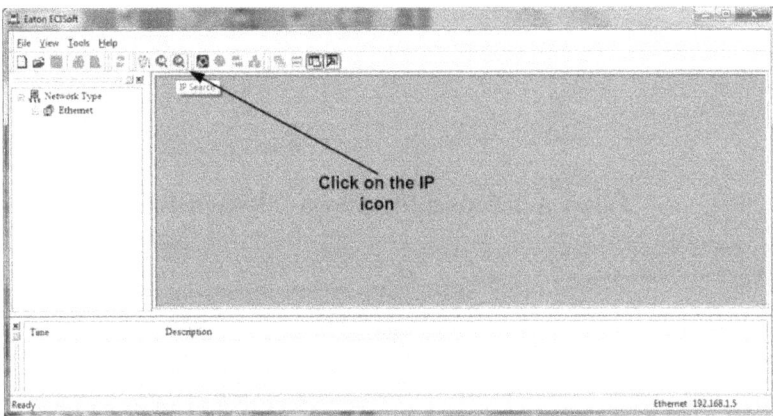

Now set an IP address of the Remote I/O to (Figure A.2):

**IP Address**: 192.168.1.2X

**Subnet mask**: 255.255.255.0

**Gateway**: 192.168.1.1

Confirm with **OK**.

**Note.** If any other device on the network has the same IP address, you need to select other address that is not in use.

If your device is not found, then you need to ensure that the communication IP address is set to 255.255.255.255 first and then scan for IP again. Follow the steps in Figure A.3

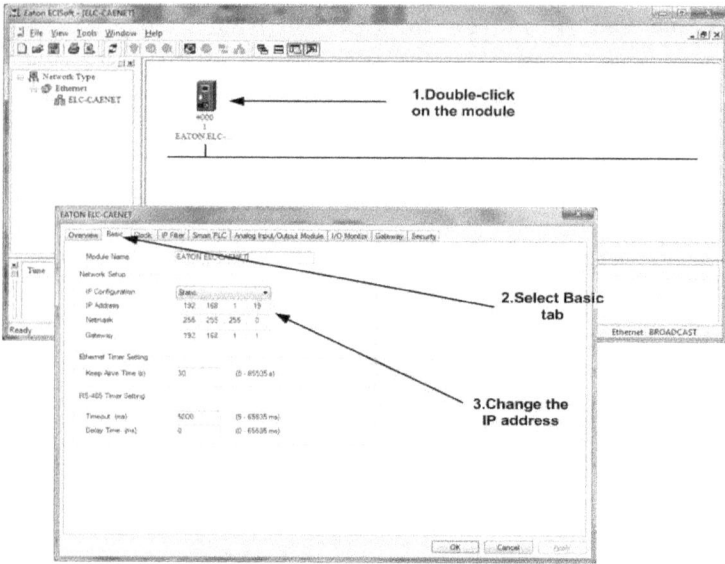

**Figure A.2:.Setting IP Address of Remote IO**

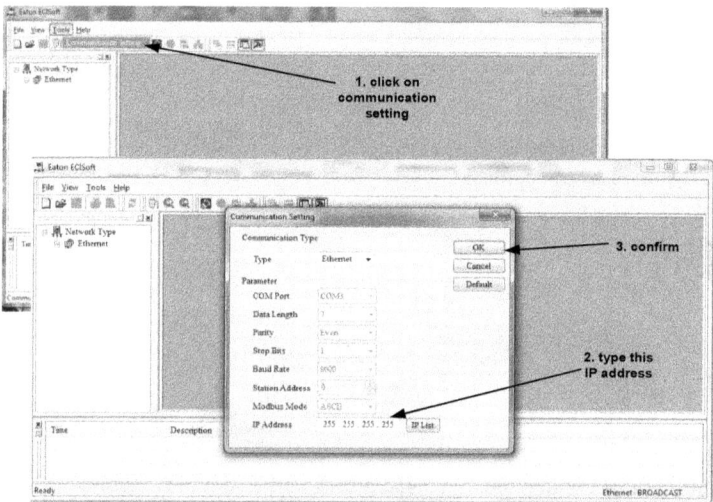

**Figure A.3: Steps to Set IP Address to 255.255.255.255**

## A.3.4 Assign and IP address to: C441 Ethernet Module of Motor Insight Overload and Monitoring Relay [1]

i) The IP address of C441 Ethernet Module of the Motor Insight is configured using the dip switch accessible on the top of the module.

Below is the information from the manual on how to set up the IP address of the C441 Ethernet Module [1].

### 4.2 Setting the IP Address Mode

Though the Ethernet Module has two Ethernet Ports, it only has one IP address that is used to target communications to the device. The dip switch accessible on the top of the module is used to establish the IP address Mode. The switch settings and the resulting behaviors are depicted in the table below.

**Table 9. IP Address Switch Settings**

| | Value | Mode | Behavior at reset (power cycle or configuration reset button) |
|---|---|---|---|
| 1<br>2<br>4<br>8<br>16<br>32<br>64<br>128<br><br>On | 0 | Restore | The operating IP configuration will be set to the follow values:<br>IP Address = 192.168.1.254<br>Net Mask = 255.255.255.0<br>Gateway = 192.168.1.1<br>**Note:** This mode is intended for fast recovery from an unknown static IP configuration. The switch Value must be changed to apply a new IP setting |
| | 1-253 | Static (HW) | The Value determines the last byte of the IP address. The rest of the IP configuration will be equal to the Static NV values set via web pages or other protocol.<br>**Note:** This mode is intended for applications where fast deployment of devices without web configuration is important. |
| | 254 | Static (NV) | The IP configuration will be set to the values stored in NV memory. The default NV values from the factory are:<br>IP Address = 192.168.1.254<br>Net Mask = 255.255.255.0<br>Gateway = 192.168.1.1<br>These can be changed from the web page or by writes to modbus registers. |
| | 255 | DHCP | The IP configuration is set by an external DHCP server on the network. |

**Note.** You can only change the last octet number of the IP address of the Module. So you cannot change the first three octet numbers **192.168.1.3X**

So the dip switches only change the last octet number. The dip switches represent a number in binary format. With all the switched turned off (switched to zero), then the default IP address is **192.168.1.254**

When using the equipment in the lab, you probably going to see the dip switches from the top view (rotate 180°) .The dip switch in the bottom represent the least significant bit.

**Example (Figure A.4):**

So to Set the IP address to 192.168.1.10

Switch on the dip switches that correspond to the numbers. In this case, the dip switches of 2 and 8 are turned on (to get number 10).

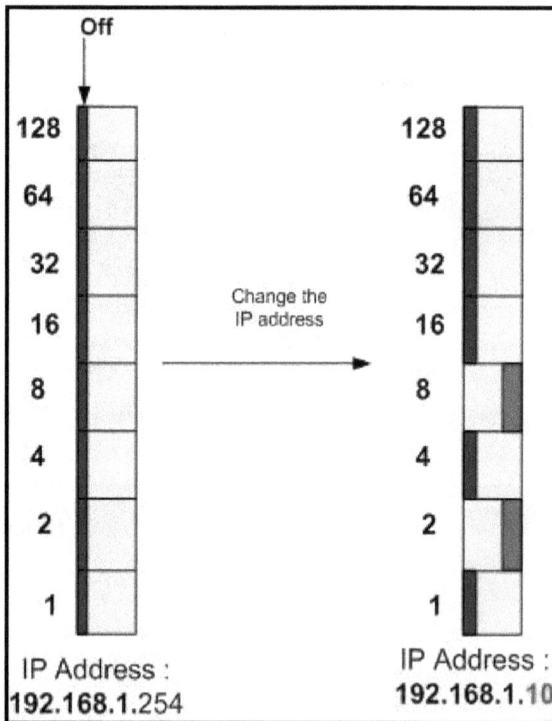

**Figure A.4: Dip Switches Top View**

Now set an IP address of the C441 Ethernet Module to:

**IP Address**: 192.168.1.3X

**Note.** You need to power off the board, and power it on again to activate the new IP address.

i)  Open Internet explorer browser and type the IP address in the address bar of the browser

ii)  Click on the Network configuration tab on the left hand side to see the network information of the module. The IP address of the device, the Subnet mast and the gateway.

**A.4    Part 2: Assign IP address using web browser**

**A.4.1    Assign and IP address to: C441 Ethernet Module of the Motor Insight**

i)  Open Internet explorer browser and type the IP address in the address bar of the browser

ii)  You can change the IP address by clicking on Network configuration tab → Stored Ethernet Address as shown below.

**Note. .**

- You can only change the last octet number when selecting a new IP **(192.168.1.X)**
- Don't change the stored Ethernet Default Gateway

You need to change the dip switches too to match the new IP address, and you need to power off the board, and power it on again to activate the new IP address.

iii)  To check if the IP address was changed, use the ping command to ping the new IP address. You can also try to access it again using the web browser.

[181]

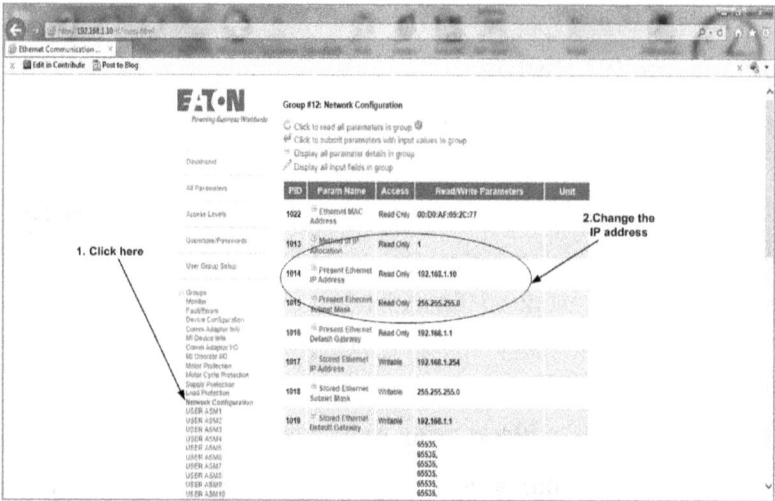

## A.4.2 Assign and IP address to DA1 (motor drive)

You can change the IP address of the Variable frequency drive using the web browser.

i) Open Internet explorer browser and type the IP address in the address bar of the browser

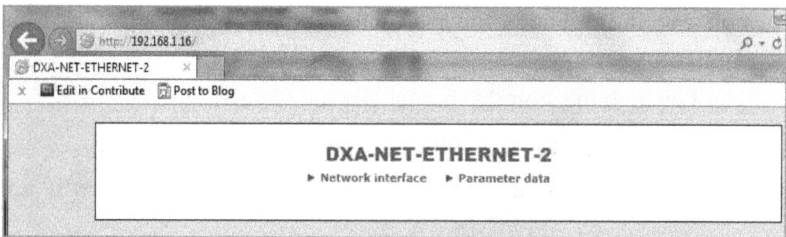

ii) Click on Network Interface → Network Configuration. And then select the new IP address

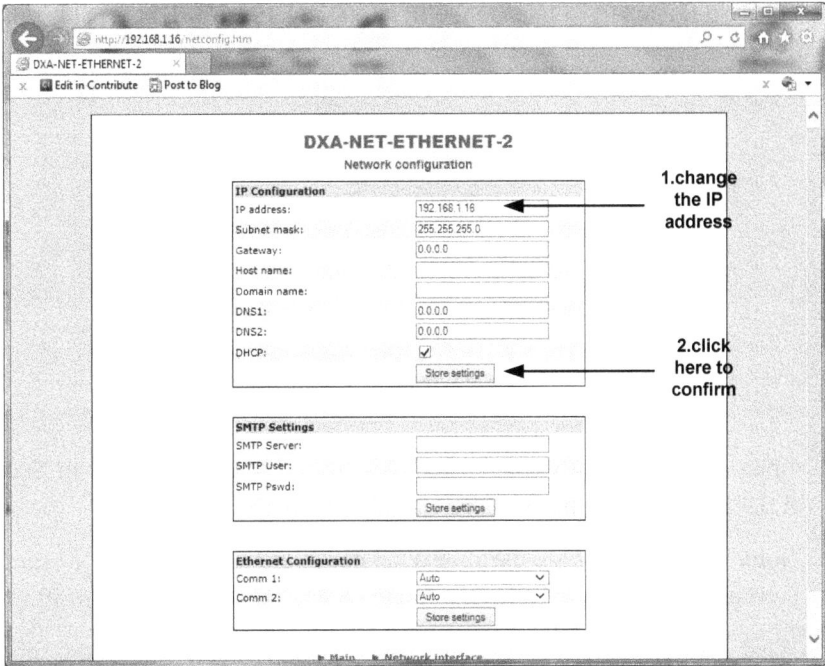

**DXA-NET-ETHERNET-2**
Network configuration

**IP Configuration**
IP address: 192.168.1.16
Subnet mask: 255.255.255.0
Gateway: 0.0.0.0
Host name:
Domain name:
DNS1: 0.0.0.0
DNS2: 0.0.0.0
DHCP: ☑
Store settings

1.change the IP address

2.click here to confirm

**SMTP Settings**
SMTP Server:
SMTP User:
SMTP Pswd:
Store settings

**Ethernet Configuration**
Comm 1: Auto
Comm 2: Auto
Store settings

▶ Main   ▶ Network Interface

**Note.** You need to power off the board, and power it on again to activate the new IP address.

iii)  Open Anybus IP config software again to check that the new IP address is assigned. Use the ping command to ping the new IP address. You can also try to access it again using the web browser.

iv)  B-Questions:

1. What data can you see when you use web browser? What options do you have?

2. Can you monitor/change the drive status/parameters?

**A.4.3   Assign and IP address to ELC Distributed I/O (ELC-CAENET) device**

You can change the IP address of the remote I/O using the web browser.

i)  Open Internet explorer browser and type the IP address in the address bar of the browser

[183]

ii)   Click on Basic tab → change the IP address to 192.168.1.21 → click apply.

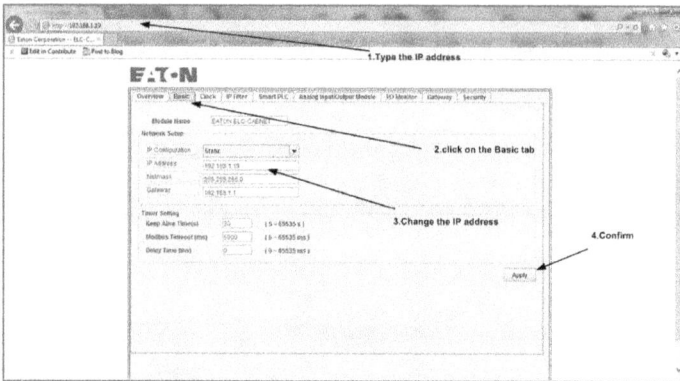

iii)  Open ECISoft software again and follow the same procedure in Section A.3.3 to check that the IP address has been changed.

**Note.** The other tabs are used for other configuration settings. For example; you can write a simple PLC program to control the I/O using Smart PLC tab.

iv)   What data can you see when you use web browser? What options/tabs do you have?

v)    Reset the IP address of your computer by turning on the DHCP.

vi)   Power off the board, disconnect the cables.

**Note:** keep your notes of this lab to use them later in your final report.

## A.5   References

[1] Eaton corporation, C441 Ethernet Module User Manual, Available as of June 30, 2014, www.eaton.com

[2] Eaton corporation, DX-NET-ETHERNET-2 Field bus connection Ethernet IP for Variable Frequency Drives DA1 User Manual, Available as of June 30, 2014, www.eaton.com

[3] Eaton corporation, ELC Distributed I/O Adapters Manual, Available as of June 30, 2014, www.eaton.com

[184]

# Appendix B

## Lab B: Ethernet IP Configuration of Communication and Data Access

### B.1 Objective

The purpose of this laboratory is to program communication among Ethernet IP devices from different vendors. The laboratory helps students to learn how to program Ethernet IP implicit UDP messaging, and Ethernet IP explicit TCP connected and unconnected messaging.

### B.2 Materials

The following materials are required for this laboratory:

i)   Eaton Ethernet IP – Smartwire Unit

ii)  Productivity 3000 PLC

iii) Personal Computer ("the PC") with Anybus IPconfig, Eaton EICSoft for the remote I/O device and Productivity Suite Programming Software

### B.3 Procedure

### B.3.1 Connections and Initial Setup

i)   Plug the RJ45 Ethernet connector into the Eaton Ethernet switch.

ii)  Plug the Eaton Ethernet IP- Smartwire unit in power. Now your network is similar to Figure 4.3, but without the productivity 3000 PLC.

iii) Turn on your computer

### B.3.2 Connect your Computer to the Lab LAN

i)   To view Network Connection of your computer click start →control panel→ network and sharing center →local area connection2→properties.

ii)  Once local area connection 2 properties window open, select "Internet Protocol Version 4 (TCP/IPv4)", then click properties (see Figure B.1).

iii) In the "Internet Protocol Version 4 (TCP/IPv4)" window, select "Use the following IP address" option and enter the following settings: **IP Address:** 192.168.1. **C, Subnet mask:** 255.255.255.0, **Gateway**: 192.168.1.1, where

PAT-**C** is your computer station number. For example if your computer station is PAT-**6**, then your IP address is <u>192.168.1.**6**</u>.

**B.3.3   Assign IP Address to Eaton VFD and to ELC-CAENET Remote** I/O

i)   Anybus IPconfig software is used to configure and assign the IP address to DA1 (Variable Frequency Drive - VFD).

ii)  To assign IP address to the VFD, Open Anybus IPconfig : by clicking Start → programs → HMS → Anybus Ipconfig.

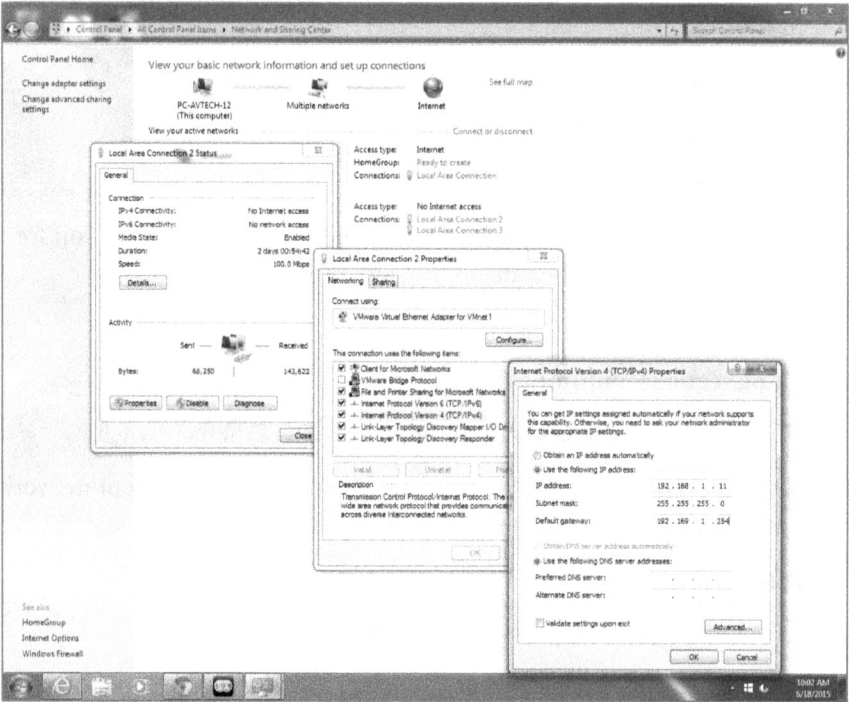

**Figure B.1: PC Network Connections Settings**

iii)  Assign the following communication parameters to the device:

- IP Address: 192.168.1. XC; X is your lab section and PAT-C is your computer station number.

- Subnet mask: 255.255.255.0.

- Gateway: 192.168.1.1.

[186]

For detailed description of the process of assigning communication parameters to DA1 – VFD refer to LAB 1A: Ethernet IP Configuration lab.

iv) To assign an IP address to the remote I/O, Eaton ECISoft software is used. Open the software by clicking start → programs → Eaton→ Communication →ECISoft and assign the following communication parameters to the device:

- IP Address: 192.168.1. (X+1)C; L0X is your lab section and PAT-C is your computer station number. Do not include brackets when entering IP address.

- Subnet mask :255.255.255.0

- Gateway: 192.168.1.1

For detailed description of the process of assigning communication parameters to the remote I/O refer to LAB 1A: Ethernet IP Configuration lab.

v) Save the screen shots on your USB memory stick to show in your report how you set the communication information.

**B.3.4 Configuration of PLC Communication Parameters**

i) Connect the Productivity 3000 PLC to your PC using a USB cable.

ii) Plug the power cable of the PLC in power.

iii) Open the Productivity Suit software and click on read Project from PAC

iv) Click on "Hardware Config" in the project tree window. New window opens.

v) Click on the image under "Local Base Group" (Figure B.2). New window in Figure B.3 opens.

vi) Click on the PLC processor highlightened by a red rectangle in Figure B.3. Window in Figure B.4 opens. Then clik on "Ethernet Ports".

vi) For "Port Security Setting", select "Read/Write; and for "Use the following:, enter the following communication information:

[187]

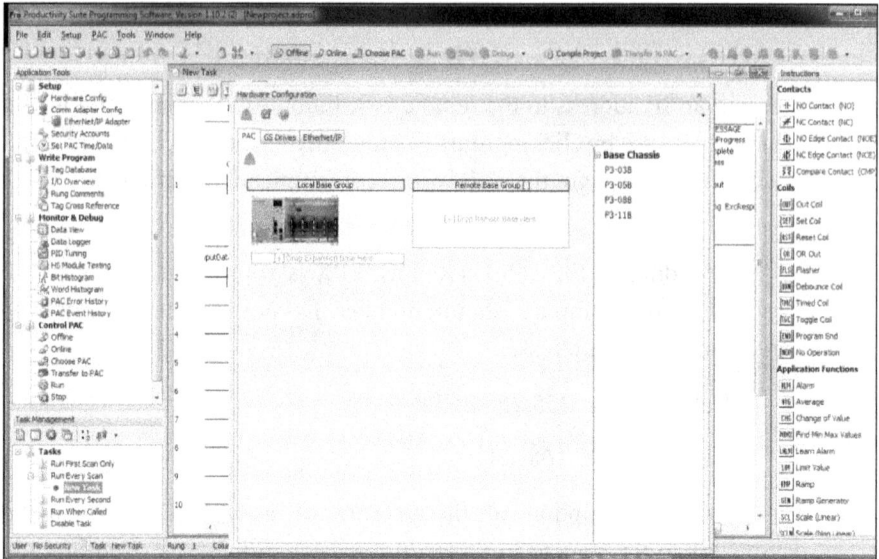

**Figure B.2: Hardware Configuration**

- IP Address: 192.168.1. (X+2)C, L0X is your lab section and PAT-C is your computer station number. Do not include brackets when entering IP address.

- Subnet mask :255.255.255.0

- Gateway: 192.168.1.1

vii) Save the screen shots on your USB memory stick to show in your report how you set the communication information.

viii) Click "OK" and close the "Hardware Configuration" screen.

[188]

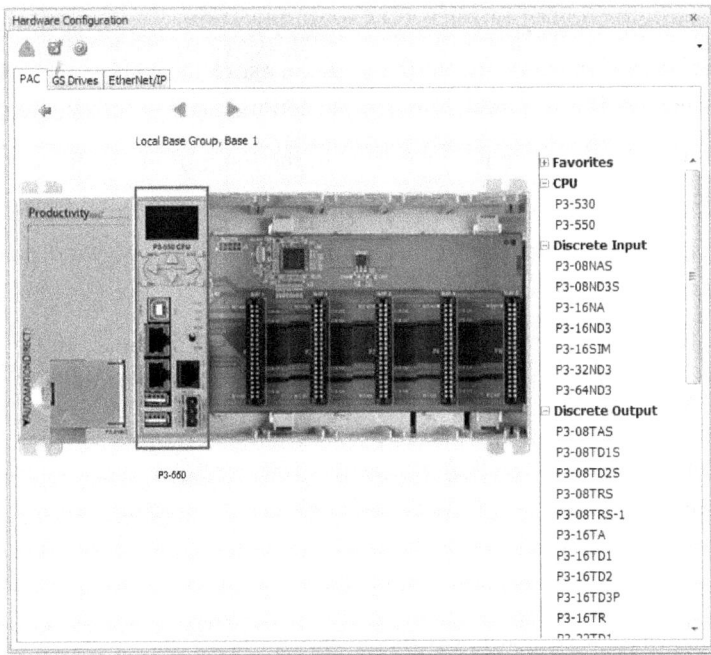

**Figure B.3: Network Configuration**

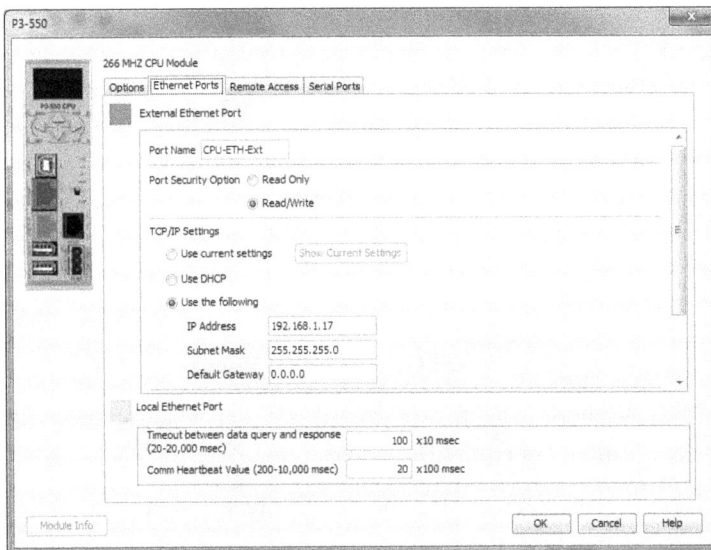

**Figure B.4: Configuration of Ethernet Port of Productivity 3000 PLC**

## B.3.5 Configuration of Implicit IO Communication

i) Save the screen shots on your USB memory stick throughout this lab to show in your report how you configured the communication information.

ii) Connect the Productivity 3000 PLC to the Eaton Ethernet Switch using an Ethernet cable.

iii) Click on "Hardware Config" in the project tree window, window in Figure B.2 opens.

iv) Click on "Ethernet/IP", new window in Figure B.5 opens. In this window there are two clients already configured. Your window should be empty.

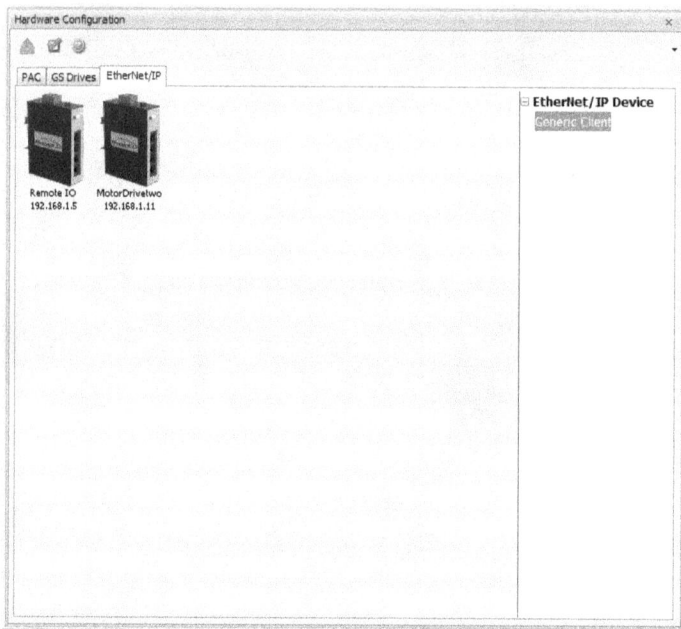

**Figure B.5: Generic Client Window**

v) Click on "Generic Client" and drag it into the window under Ethernet/IP heading. Window in Figure B.6 opens.

vi) In "Device Name", enter "RemoteIOXC", where X is your lab section and C is your computer station.

vii) Enter names for "TCP Connected", Adapter Name", "Vendor ID", and "TCP Error". Every time you enter a tag you are reminded that the tag needs to be created through the window in Figure B.7. Click "ok" to create the tag.

viii) IP Address is the IP address you configured into the Eaton Remote IO.

ix) Click on the green + sign in Figure B.6 and select "Add IO Message". The window expands to look as shown in Figure B.8.

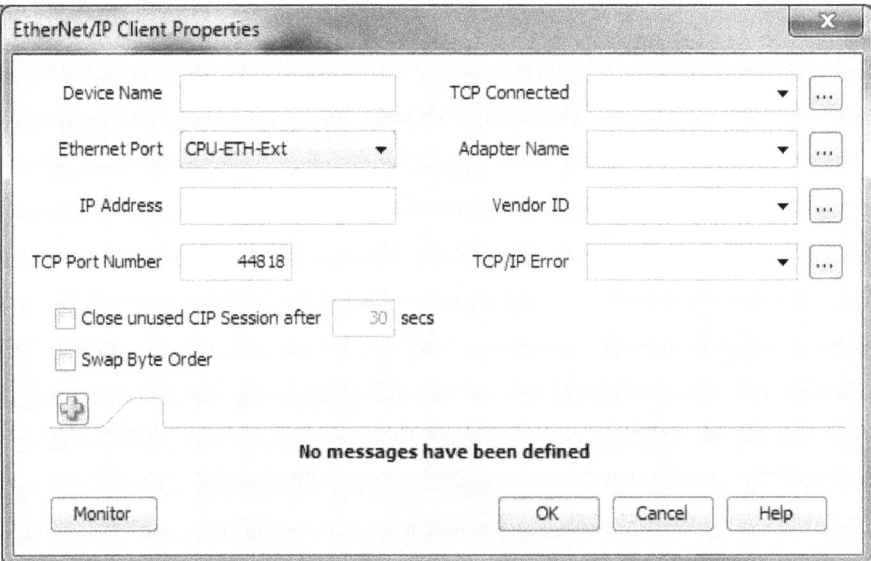

**Figure B.6: Ethernet IP Client Communication Parameters**

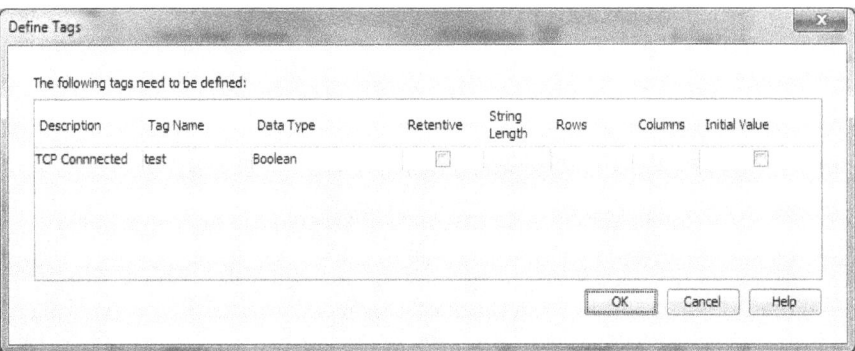

**Figure B.7: Tag Definition**

[191]

EtherNet/IP Client Properties

| | | | | | |
|---|---|---|---|---|---|
| Device Name | | TCP Connected | | ▼ | ... |
| Ethernet Port | CPU-ETH-Ext ▼ | Adapter Name | | ▼ | ... |
| IP Address | | Vendor ID | | ▼ | ... |
| TCP Port Number | 44818 | TCP/IP Error | | ▼ | ... |

☐ Close unused CIP Session after 30 secs
☐ Swap Byte Order

IO MSG 1

| | | | | | |
|---|---|---|---|---|---|
| Enable | | ▼ ... | Connection Online | | ▼ ... |
| | | | General Status | | ▼ ... |
| ☐ Enable Routing | Slot Number | 0 | Extended Status | | ▼ ... |
| | | | Status Description | | ▼ ... |

T->O (INPUT) | O->T (OUTPUT) | CONFIG DATA

Target To Originator (INPUT) Data

| | |
|---|---|
| Delivery Option | Multicast ▼ |
| RPI Time (msec) | 250 |
| Connection Point | 0 ( 0x0 ) |
| Datatype: | ----- |
| Data Array | ▼ ... |
| Message Size (bytes): | 0 |
| Number of Elements | 0 |

Monitor        OK   Cancel   Help

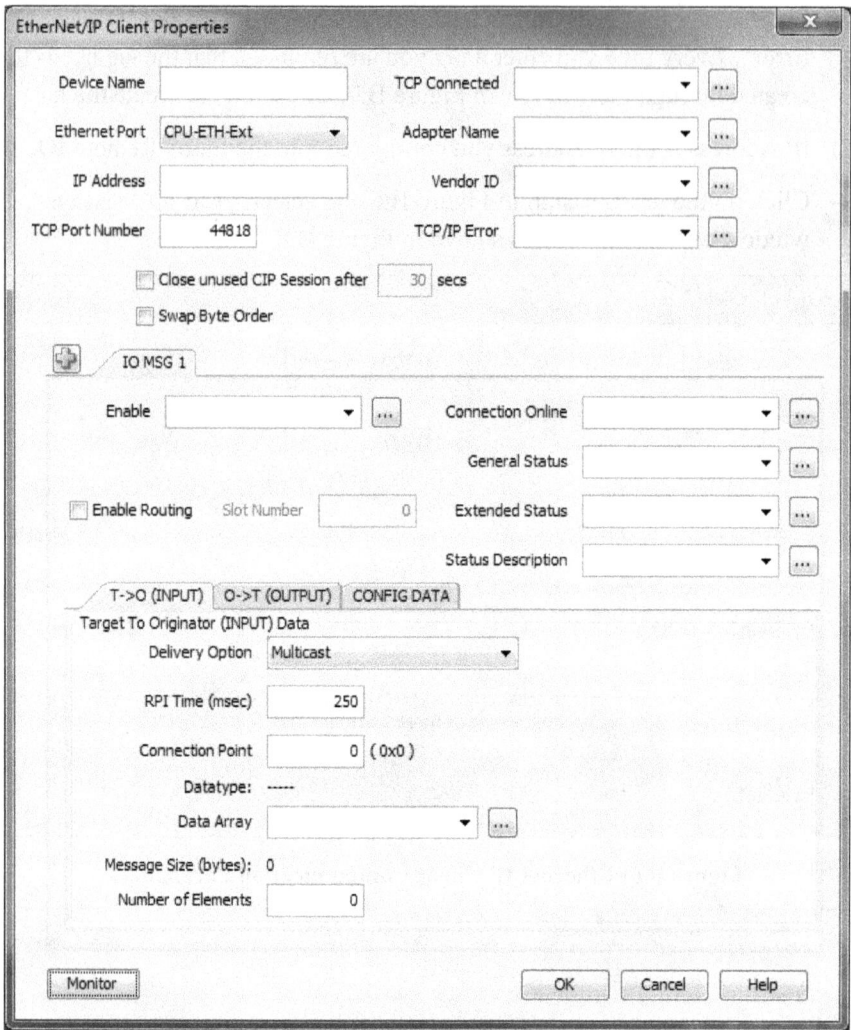

**Figure B.8: Expanded Ethernet IP Client Properties**

x) Click on "T>O (INPUT)" and set the "Delivery Option" to "Multicast".

xi) Leave the data update rate "RPI Time" at 250msec. Set the "Connection Point" to 104 and the" Number of Elements" to 24 (Figure B.10). These parameters can be found in the device manual, and they are explained in Section 4.2.1.

[192]

xii) Your data array should be "RemoteIOInputDataXC" X is your lab section and C is your computer station. You will get a message that you need to increase the size of the data array. Enter a value of 25 in the window shown in Figure B.9, that open when you click "OK".

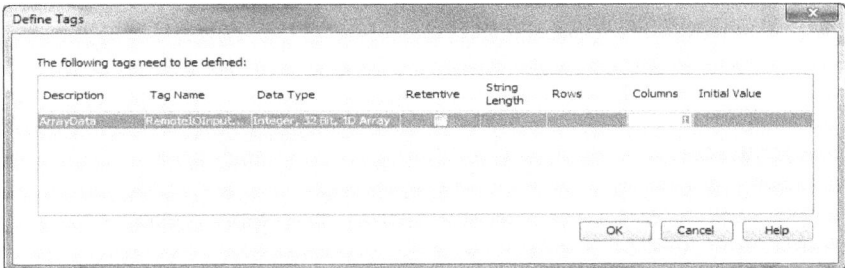

**Figure B.9: Setting Array Size**

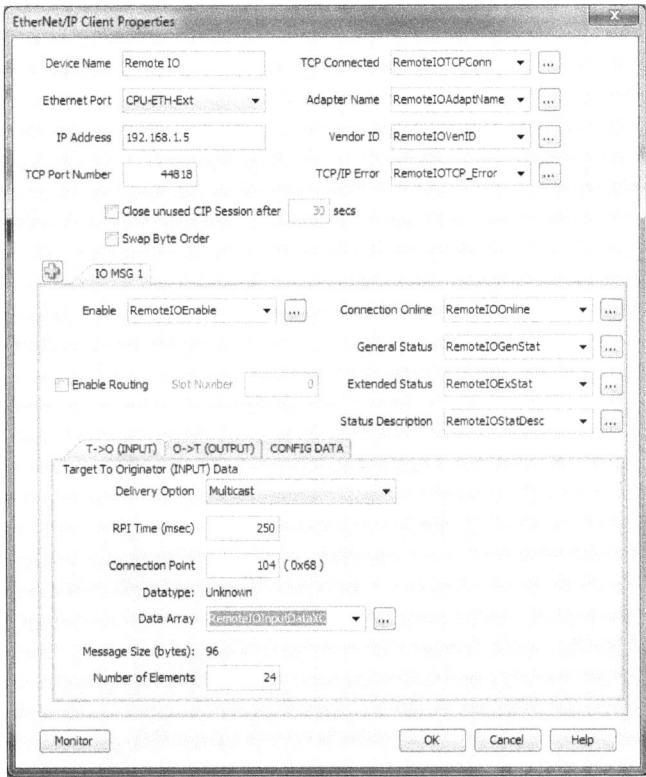

**Figure B.10: Output Parameters**

xiii) Click on "O>T (OUTPUT)"

xiv) Leave the data update rate "RPI Time" at 250msec. Set the "Connection Point" to 105 and the "Number of Elements" to 20 (Figure B.10). These parameters can be found in the device manual, and they are explained in the lab support YouTube video.

xv) Your data array should be "RemoteIOOutputDataXC" X is your lab section and C is your computer station. You will get a message that you need to increase that size of the data array. Enter a value of 20 in the window shown in Figure B.9, that open when you click "OK".

xvi) Click on "CONFIGDATA"

xvii) Set the "Connection Point" to 50 and the "Number of Elements" to 0 (see Figure B.10). These parameters can be found in the device manual, and Section 4.2.1.

xviii) Your data array should be "RemoteIOConfigDataXC" X is your lab section and C is your computer station.

xix) Click on "Monitor" in Figure B.10, so that you can be able to monitor the tags associated with this client. Window in Figure B.11 opens and click "OK".

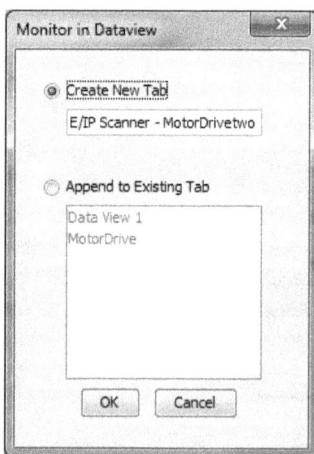

**Figure B.11: Monitor File for the Remote IO Tags**

xx) Click "OK" in Figure B.10.

xxi) Click on "Online", then "Transfer to PAC" in the main window.

xxii) After transferring the program, Click on "Data View" in the project tree window, window in Figure B.12 opens

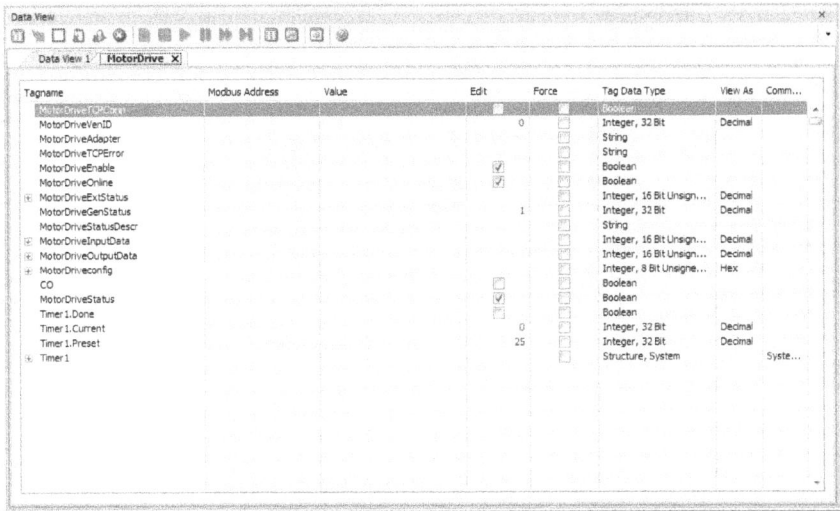

| Tagname | Modbus Address | Value | Edit | Force | Tag Data Type | View As | Comm... |
|---|---|---|---|---|---|---|---|
| MotorDriveTCPClconn | | | | | Boolean | | |
| MotorDriveVenID | | 0 | | | Integer, 32 Bit | Decimal | |
| MotorDriveAdapter | | | | | String | | |
| MotorDriveTCPError | | | | | String | | |
| MotorDriveEnable | | | ✓ | | Boolean | | |
| MotorDriveOnline | | | ✓ | | Boolean | | |
| MotorDriveExtStatus | | | | | Integer, 16 Bit Unsign... | Decimal | |
| MotorDriveGenStatus | | 1 | | | Integer, 32 Bit | Decimal | |
| MotorDriveStatusDescr | | | | | String | | |
| MotorDriveInputData | | | | | Integer, 16 Bit Unsign... | Decimal | |
| MotorDriveOutputData | | | | | Integer, 16 Bit Unsign... | Decimal | |
| MotorDriveconfig | | | | | Integer, 8 Bit Unsigne... | Hex | |
| CO | | | | | Boolean | | |
| MotorDriveStatus | | | ✓ | | Boolean | | |
| Timer1.Done | | | | | Boolean | | |
| Timer1.Current | | 0 | | | Integer, 32 Bit | Decimal | |
| Timer1.Preset | | 25 | | | Integer, 32 Bit | Decimal | |
| Timer1 | | | | | Structure, System | | Syste... |

**Figure B.12: Data View**

xxiii) Click on the name of your device and check the box against the "Enable" tag. You can click on the + sign besides data tags to view the status of the various tags.

xxiv) Identify the various data tags associated with the output terminals on the remote IO by typing 1 in the data area of the tags and examining the lights on the remote IO.

xxv) You can now use the remote IO data in you ladder logic program.

**B.3.6 Configuration of Explicit Unconnected Ethernet IP Communication**

i) Save the screen shots on your USB memory stick throughout this lab to show in your report how you configured the communication information.

ii) Go to step B.3.5 iii) and go through the steps up to B.3.5 viii), in this case set the device name to VFDXC, where X is your lab section and C is your computer station. In "IP Address", enter the IP address if the VFD.

iii) Close the "Hardware Configuration" window.

[195]

iv) Click on "End" of the first rung in the programming window, and then click on the "Ethernet/IP Explicit Message" instruction. Window in Figure B.13 opens.

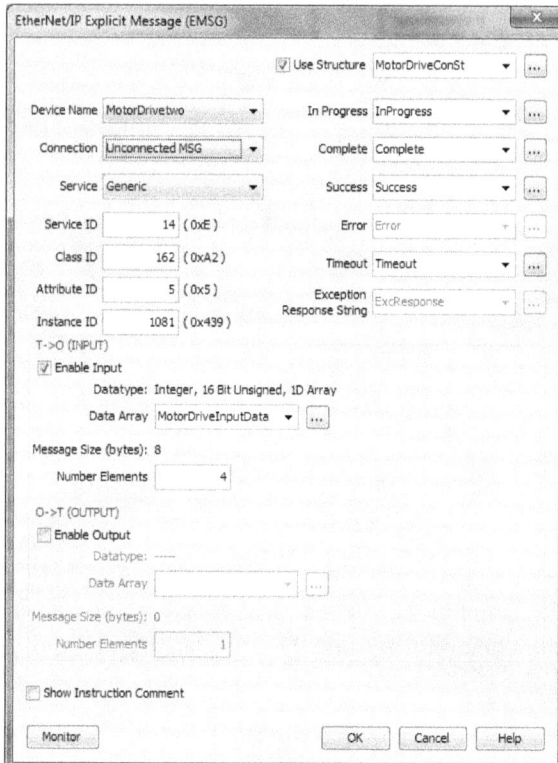

**Figure B.13: Ethernet IP Explicit Message Configuration**

v) For "Device Name", select "VFDCX"; "Connection", leave unchanged; "Service", select "Generic".

vi) Enter the following parameters, "Service ID = 14", Class ID = 162", "Attribute ID = 5", "Instance ID = 1081. These parameters can be found in the device manual, and they are explained in Section 4.2.1.

vii) Check "Enable Input", enter the name of your data array, and set the "Number Elements" to 4.

viii) Click "Monitor", then "Ok" to create monitored data, then click "Ok" to close the Ethernet IP Explicit Message configuration.

[196]

ix) The Message instruction is executed only when the rung status changes from false to true. If you want to continuously update data, you need to use a timer instruction as shown in Figure B.14.

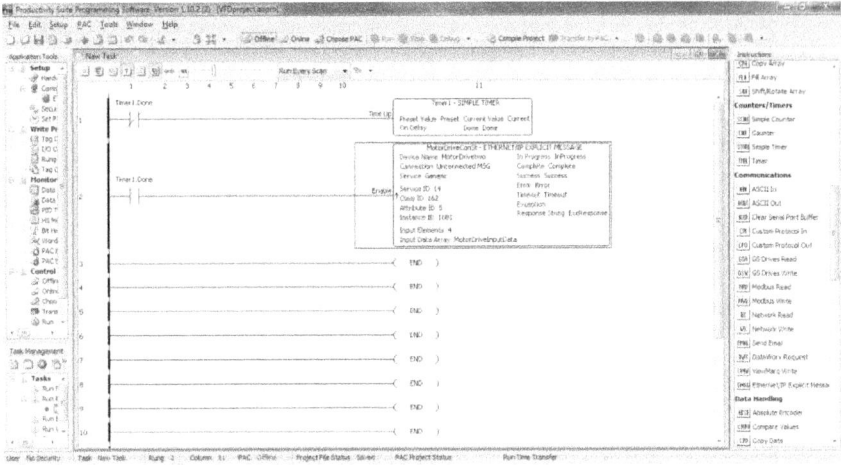

**Figure B.14: Explicit Messaging Logic**

x) Note that the *time.reset* value is entered by going to "Data View" and creating the monitoring data set.

xi) Go to "Data View" to monitor the data from the motor, and then turn on the motor. Change the motor speed to see how the parameter varies with the speed.

xii) Through the logic, scale the motor speed parameter such that the maximum speed is 60Hz and minimum speed is 0Hz.

xiii) You can now use the VFD speed data in you ladder logic program.

### B.3.7 Ladder Logic Involving Remote IO and VFD

i) Save the screen shots on your USB memory stick throughout this lab section to show in your report how your logic.

ii) Modify the logic so that if the speed is greater than **XC,** where X is your lab section and C is your computer station, a digital output goes "HIGH".

iii) Demonstrate your results to the instructor.

# Appendix C

## Lab C: Configuration of IEC61850 Device and Programming Laboratory

### C.1 Objective

The main objective of this lab is to learn how to configure IEC61850, electrical systems monitoring, protection and control devices. To demonstrate the achievement of this objective, record the data (including screen shots) that shall help you to explain your observations. In addition, include in the report the circuit diagram (including the load (lights) circuit) of the SEL Relay unit and explain how the circuit diagram is related to the relay setting in Figure C.2.

### C.2 Materials:

The following materials are required for this laboratory:

i)     SEL-751 Feeder Protection Relay Unit
ii)    Personal Computer ("the PC") with AcSELerator QuickSet and Modbus Poll software
iii)   SEL C662/C663 USB to UART Bridge
iv)    RJ-45 cable (Ethernet cable)

### C.3 Procedure

#### C.3.1 Connections

i)     Connect the SEL-C662 cable to the PC. The cable is connected to Port 3 of the SEL Relay.

ii)    Plug the SEL 751A Relay unit in power.

#### C.3.2 Run the AcSELerator QuickSet Application

i)     Locate and run the executable file for the AcSELerator QuickSet application.

ii)    On the menu bar select communications and then, on the drop down menu, select parameters.

iii)   Select Serial as the "Active Connection Type".

iv)    Select the "Serial" tab, and in the device drop down menu, select the communications port" SEL USB to UART Bridge Controller".

v) Make sure that the data speed is set to auto detect, data bits is set to 8, stop bits is set to 1 and parity is set to none. RTS/CTS should be off, DTR and RTS should be off, and XON/OFF should be on (Figure C.1); do not change any passwords.

vi) Click ok to establish a communication with the SEL device.

### C.3.3 Read the settings of the device

i) Click on the (small green) Read icon to read the settings from the connected device and then select all groups and classes to read and click ok. The program will proceed to read all 17 settings files from the memory of the SEL device.

ii) When prompted, select open files in a new editor.

iii) When the reading is completed, the left panel displays the selected directory or subdirectory of settings and when one is selected, it will display the associated settings in the right panel (Figure C.2).

### C.3.4 Configuration and Testing

i) Click on "Group 1", followed by "Set 1" and "Main". Then set the following parameters to 1: Current Transformer Turns Ration CTRN, Potential Transformer Turns Ration PTR, and Potential Transformer Synchronous Voltage Turns Ration PTRS (Figure C.2).

This must be done to program the relay for direct connections, indicating that the transformers turns ratios are 1:1 (there are no current transformers or potential transformers required when inputs are within the limitations of the terminals).

ii) Set Nominal Line Voltage VNOM to 230, Transformer Connection to WYE, and Single Phase Input Voltage SINGLEV to Y (Figure C.2).

iii) Click on "Send Active Settings" to load the setting into the relay (Figure C.3).

iv) Click on "Tools", followed by "HMI", and "HMI". Make sure both lights are off and record the voltage and current of phases A, B, and C. Also record the power, the apparent power, the reactive power, and the power factor.

[199]

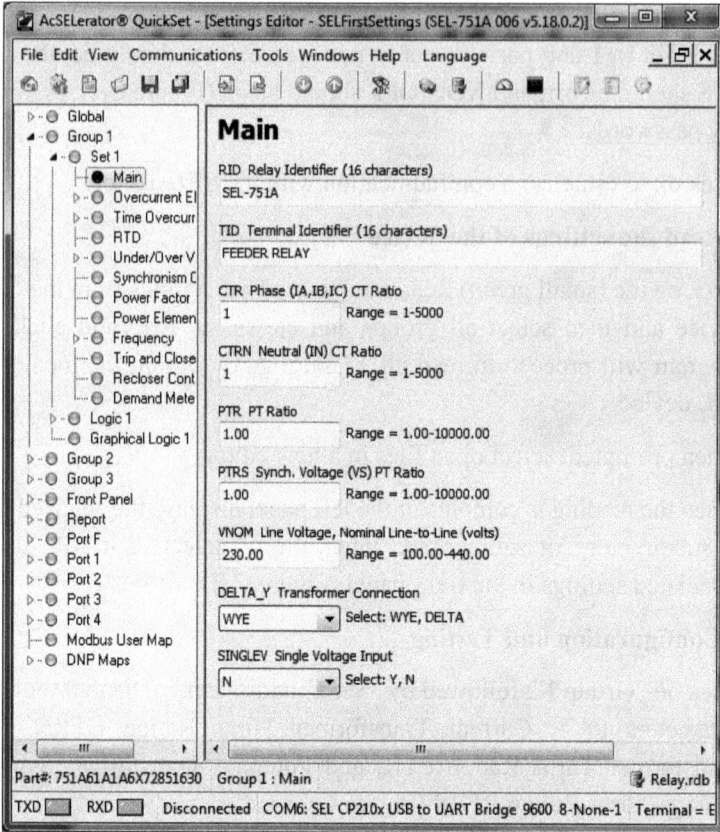

**Figure C.1: SEL Relay Metering Settings**

v) Turn on the light switch, adjust to it to full and record the voltage and current of phases A, B, and C. Also record the power, the apparent power, and the reactive power.

vi) Adjust the light switch to other positions and note what happens to the voltage and current of phases A, B, and C. As well as the power, the apparent power, and the reactive power.

vii) Close the HMI and return to the configurations screen.

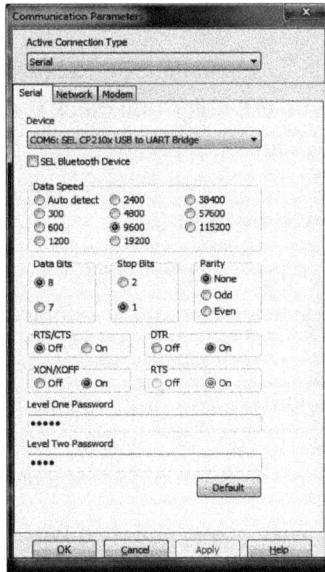

**Figure C.2: Port 3 Settings**

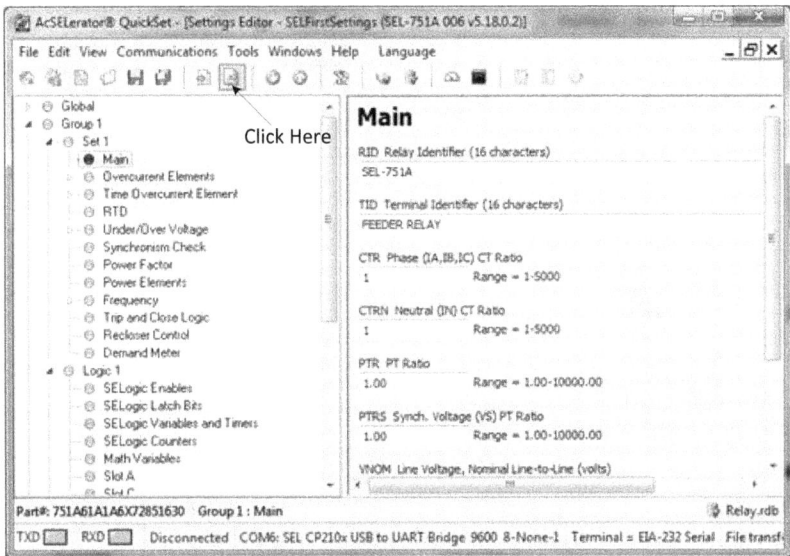

**Figure C.3: Send Active Settings**

## C.3.5 Programming the SEL 751A Relay Logic

i) Now you are going to program the SEL Relay to turn on Output 1 ("OUT301 (SELogic))". OUT301 is Output 1 on block C on the SEL Relay (I/O blocks are identified on the side of the relay). The SEL Relay unit uses this relay output to supplies 24 VDC to an external relay that controls Lamp 1. Click on "Logic 1", followed by "Slot C.

ii) Delete the content under "OUT301 (SELogic)", then click on the open icon next to the data entry space to entry new content (Figure C.4). New window in Figure C.5 opens.

iii) Open the "Analog Quantities" drop menu, then click on "instantaneous Metering", followed by double clicking on VA_MAG. This selects the instantaneous value of the magnitude of the phase A voltage.

iv) Click on the "Less Than" sign, and then type 80. The entry in the window should be VA_MAG<80 (Figure C.5). Click "Accept".

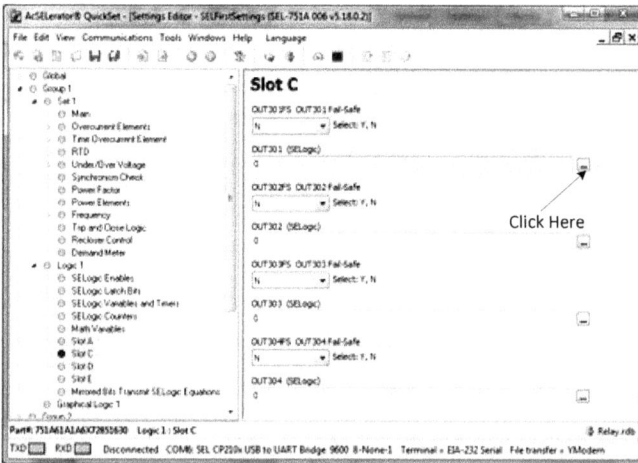

**Figure C.4: SLOT C Logic Window**

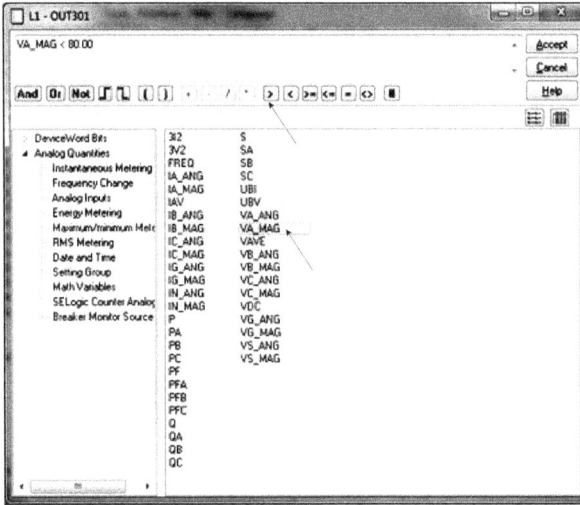

Figure C.5: Output 1, Block C Logic Editing Window

v) The configuration window looks as show in Figure C.6. This entry means that Output 1 on block C is turned on if the phase A voltage is less than 80VAC.

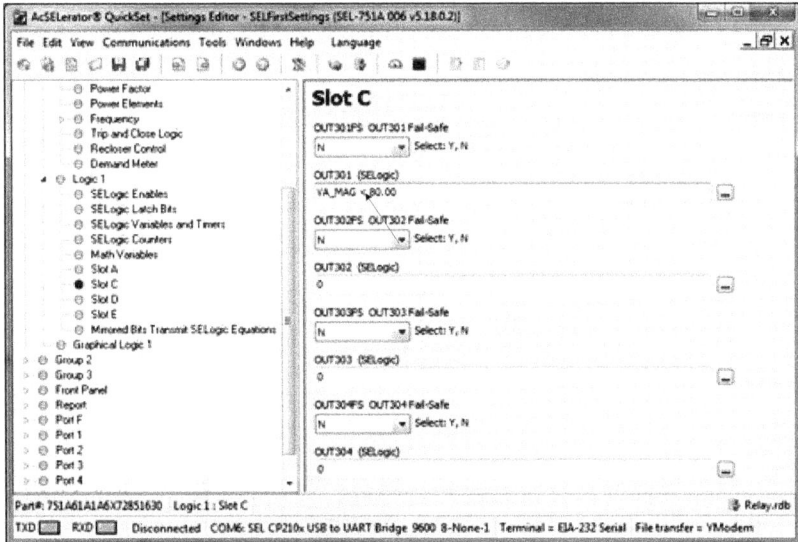

Figure C.6: SLOT C Logic Window with OUT301 Edited

[203]

vi) Click on "Send Active Settings" to load the setting into the relay.

vii) Click on "Tools", followed by "HMI", and "HMI". Make sure both lights are off and record the voltage and current of phases A, B, and C. Also record the power, the apparent power, the reactive power, and the power factor.

viii) Turn on the light switch, adjust to it to full and record the voltage and current of phases A, B, and C. Also record the power, the apparent power, and the reactive power.

ix) Observe what happens to lamp 2 as you adjust the intensity of Lamp 1.

x) Adjust the light switch to other positions and note what happens to the voltage and current of phases A, B, and C, as well as the power, the apparent power, and the reactive power.

xi) Close the HMI and return to the configurations screen.

**C.3.6 Accessing the SEL 751A IEC61850 Data using Modbus Communication**

i) Select "Port 1" on the left panel and enter the following information: "IP Address" 192.168.1.XY, "Subnet Mask" 255.255.255.0, and "Default Router Gateway" 192.168.1.254. Where X is your Lab Section and Y is your group number (Figure C.7).

ii) Select port A as the "Primary Network Port" and make sure it is in the "switched" operating mode. Enable 1 Modbus session and make sure the standard port is used (502) (Figure C.8).

iii) Send the changed configuration file by selecting the menu item with a green arrow exiting a page, or by selecting "Send..." from the file drop down menu. Select "OK" to write the file to the SEL-751A.

iv) Connect the network Ethernet cable to port A of the Relay.

v) Run the Modbus Poll program.

vi) Press "F1"on the keyboard or on the menu bar, mouse click on "Connection" and then, from the drop-down menu, select "Connect..." Select a "Modbus TCP/IP" connection.

vii) Enter the IP address of the relay. With proper configuration, the program window should display unnamed variables with their corresponding values.

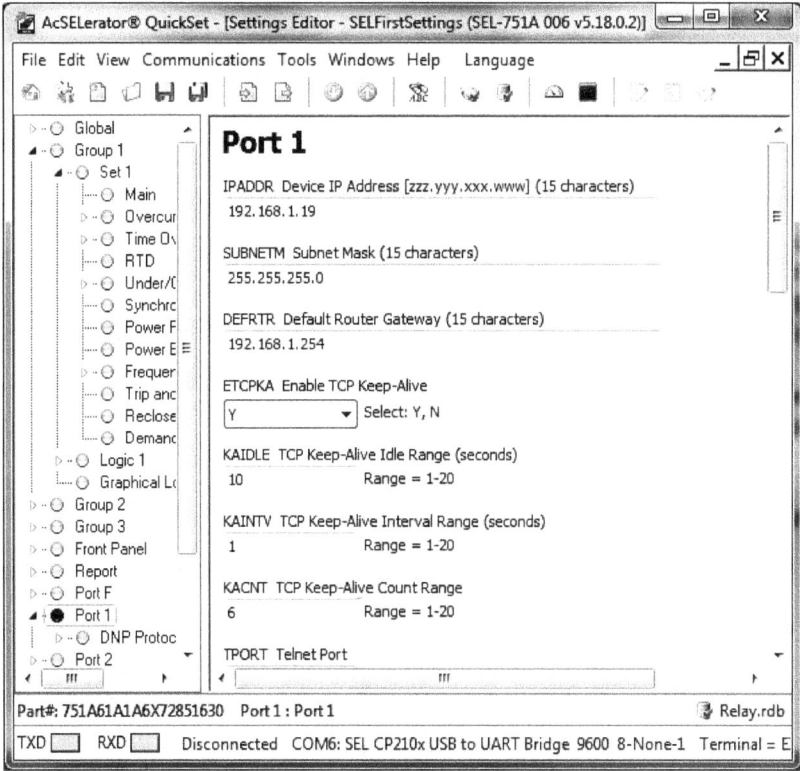

**Figure C.7: Ethernet Setting**

viii) Select "Setup" on the menu bar to access the "Read/Write Definition…", or press F8. With "Slave ID" set to "1", set the Function to "03 Read Holding Registers" Then select the starting "Address" and "Quantity" to a range that allows you to obtain the values below, which can be found in Appendix E of the SEL-751A instruction manual.

| | | | |
|---|---|---|---|
| IA_MAG | _____ | VA_MAG | _____ |
| IA_ANG | _____ | VA_ANG | _____ |
| IB_MAG | _____ | VB_MAG | _____ |
| IB_ANG | _____ | VB_ANG | _____ |
| IC_MAG | _____ | VC_MAG | _____ |
| IC_ANG | _____ | VC_ANG | _____ |

[205]

| | | |
|---|---|---|
| IN_MAG _____ | VG_MAG | _____ |
| IN_ANG _____ | VG_ANG | _____ |
| IAV _____ | VAVE | _____ |

ix) Turn on Lamp 1 and record the values of the above parameters.

x) Save your configuration on a personal memory stick.

xi) Close the AcSelerator QuickSet Application.

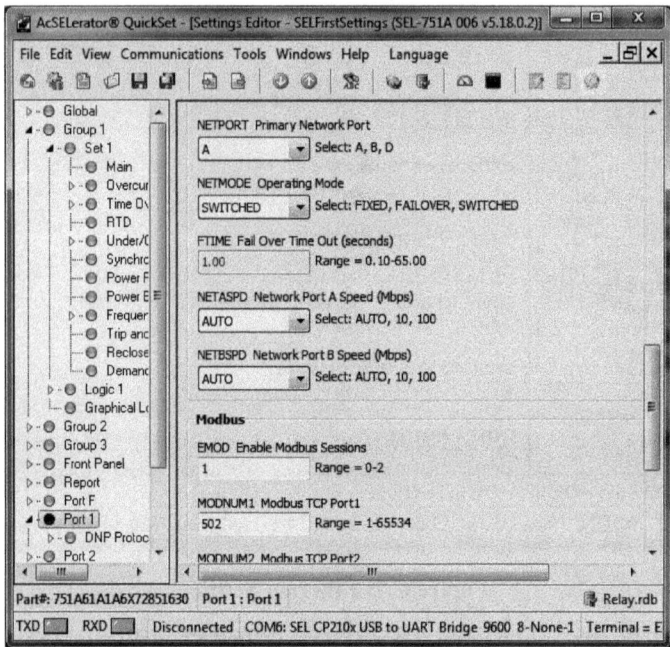

**Figure C.8: Modbus Settings**

# Appendix D

## Lab D: IEC 61850 GOOSE Messaging Laboratory

### D.1 Objective

The main objective of this laboratory is to learn how to configure, send and receive IEC 61850 GOOSE messages. To demonstrate the achievement of this objective, record the data (including screen shots) that shall help you to explain your observations.

### D.2 Materials

The following materials are required for this laboratory:

i)   SEL-751 Feeder Protection Relay Unit
ii)  Personal Computer ("the PC") with AcSELerator QuickSet (Engineering Tool) Software and AcSELerator Architect (Engineering Tool) Software
iii) SEL C662/C663 USB to UART Bridge
iv)  RJ-45 cable (Ethernet cable)

### D.3 Procedure

#### D.3.1 Connections

i)   Connect the SEL-C662 cable to the PC. The cable is connected to Port 3 of the SEL Relay.

ii)  Plug the RJ45 Ethernet connector into Port 1A of the SEL Relay.

iii) Plug the SEL 751A Relay unit in power.

#### D.3.2 Configure and Program the SEL 751A Relay

The following steps are described in detail in the SEL Relay configuration laboratory. Please refer to that laboratory handout in case of any doubt.

i)   Locate and run the executable file for the AcSELerator QuickSet application.

ii)  Open the configuration file that you saved during the SEL Relay configuration laboratory.

iii) Connect the PC to the relay through the SEL USB (Port 3).

[207]

iv) Select "Port 1" on the left panel and enter the following information: "IP Address" 192.168.1.XY, "Subnet Mask" 255.255.255.0, and "Default Router Gateway" 192.168.1.254. Where X is your Lab Section and Y is your group number.

v) Now you are going to program the SEL Relay to turn on Output 2 ("OUT302 (SELogic))". OUT302 is Output 2 on block C on the SEL Relay (I/O blocks are identified on the side of the relay). The SEL Relay unit uses this relay output to supplies 24 VDC to an external relay that controls Lamp 2. Click on "Logic 1", followed by "Slot C.

vi) Delete the content under "OUT302 (SELogic)", then click on the open icon next to the data entry space to enter new content (Figure D.1). New window in Figure 3 opens.

vii) Open the "DeviceWord Bits" drop menu, then click on "Virtual Bits", followed by double clicking on VB001(Figure D.2). Click "Accept".

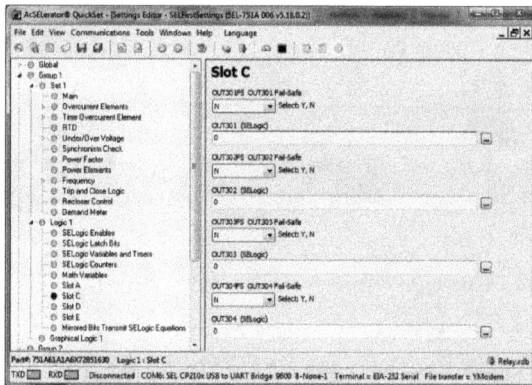

**Figure D.1: SLOT C Logic Window**

[208]

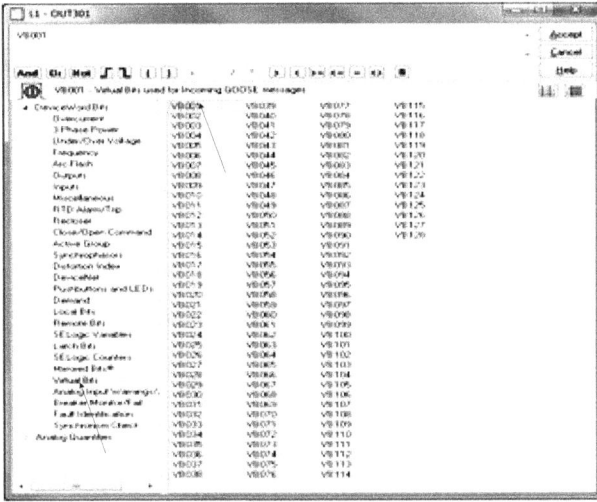

**Figure D.2: Output 1, Block C Logic Editing Window**

viii) The configuration window now looks as show in Figure D.3, but for this case VB001 should be in OUT302 slot. This entry means that Output 2 on block C is turned on when the received GOOSE bit assigned to Virtual Bit 1 (VB001) is HIGH.

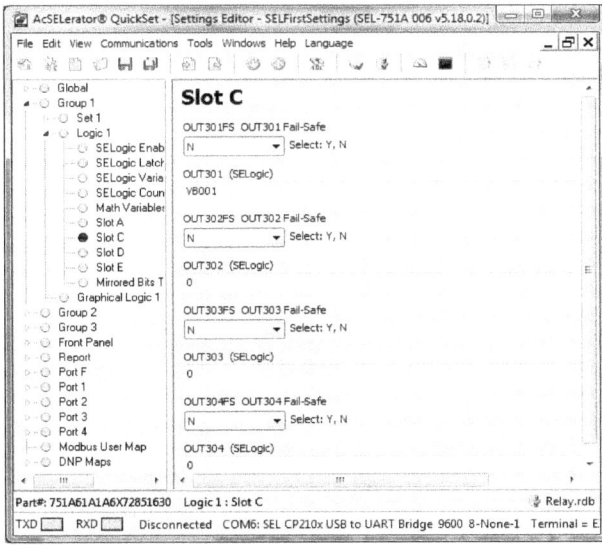

**Figure D.3: SLOT C Logic Window with OUT301 Edited**

[209]

ix)    Click on "Send Active Settings" to load the setting into the relay.

Note that it is possible to use "Graphic Logic Editor" to program the SEL Relay (Figure D.4)

### D.3.3    Configure SEL 751A Relay GOOSE Message

i)     Locate and run the executable file for the AcSELerator Architect application.

ii)    Click on 'SEL LAB" on desktop, open the AcSELerator Architect folder and then double click on the ".selaprj" file

iii)   The GOOSE message transmitter is already configured in this file. It is for the device "SEL_751A_2" which sends out a status variable of 0 if the voltage of Phase A of the motor that the IED monitors is less than 80VAC and a 1 if the voltage is greater than 80VAC.

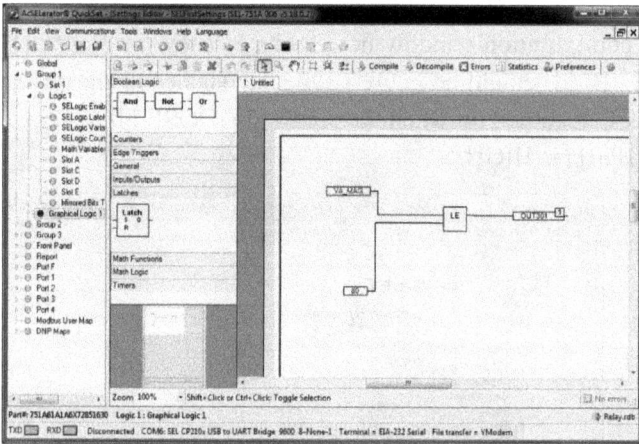

**Figure D.4: SEL Relay Graphical Logic Editor**

iv)    SAVE THIS FILE ON YOUR MEMORY STICK BEFORE CONTINUEING THE LABORATORY.

v)     Click on "GOOSE Transmit" to see the GOOSE message the device sends. Double click on the message to see its configuration and its associated dataset (Figure D.5).

vi)    DO NOT CHANGE ANYTHING IN THIS CONFIGURATION.

[210]

vii) Right click on "SELRelay1", move cursor over "Add IDE" then double click "SEL_751A" A. New window opens.

viii) Select IED of "Class File Version = 002", and "Description = R300 or greater" (Figure D.6).

ix) A new IED is added with default Name "SEL_751A_1". Right click on it and rename it "SEL_751A_1_XY, where X is your lab section and Y is your lab group number.

x) Click on "Properties" and enter the device IP address, subnet mask, and gateway. These are the setting you entered for Port 1 in AcSELerator QuickSet.

xi) Click on "GOOSE Receive". Now you can see the GOOSE message(s) your device can subscribe to under the "GOOSE Receive" column. Open the message to reveal its data by clicking on the + icons (Figure D.7).

xii) This data represents the status of output 1 and 2 on block C of the GOOSE transmitting device.

xiii) Drag the stVal (Status Value) under Ind01 to the "subscribed Data Items" column, and in front of "VB001". This means that your device shall receive the status value of Output I of the transmitting device and it shall assign it to internal (Virtual) tag (bit) VB001. You can now use this tag to take any action you want. For this lab, you have already configured your device to control output 2 on block C using VB001 in Section D.3.2.

xiv) We do not want your device to send any GOOSE messages. Therefore, click on "GOOSE Transmit", right click on the GOOSE message in the "GOOSE Transmit" column, and delete it.

xv) To transfer the configuration to the SEL relay, right click on your IED "SEL_751A_1_XY", and then click "send CID". Window in Figure D.8 opens. FTP address is the IP address of your device, User Name is FTPUSER, and Password is TAIL.

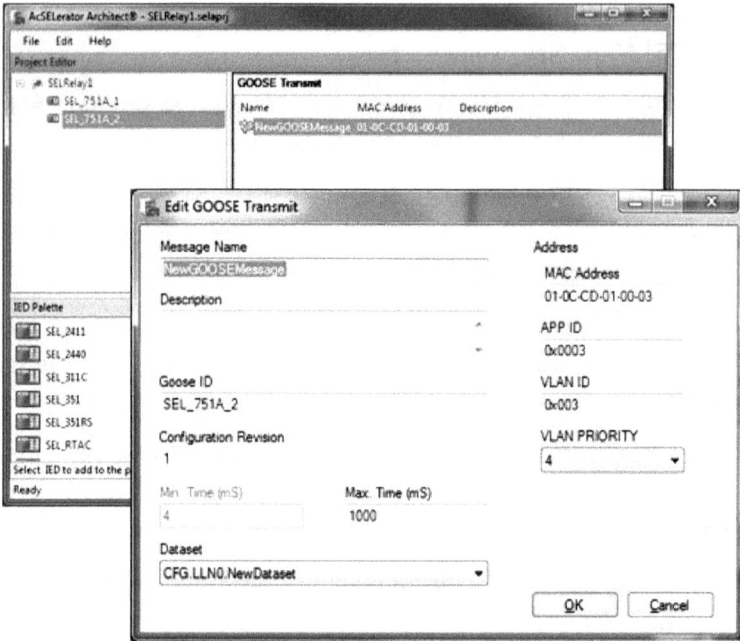

**Figure D.5: GOOSE Message Configuration**

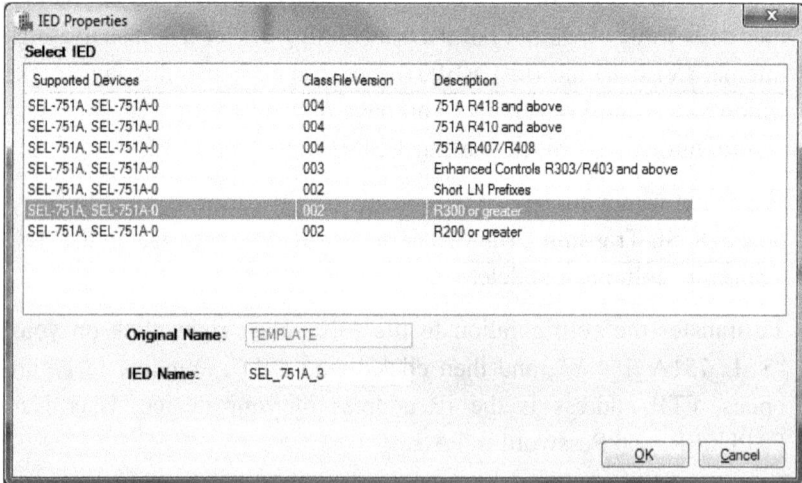

**Figure D.6: Adding GOOSE Messages Configuration File**

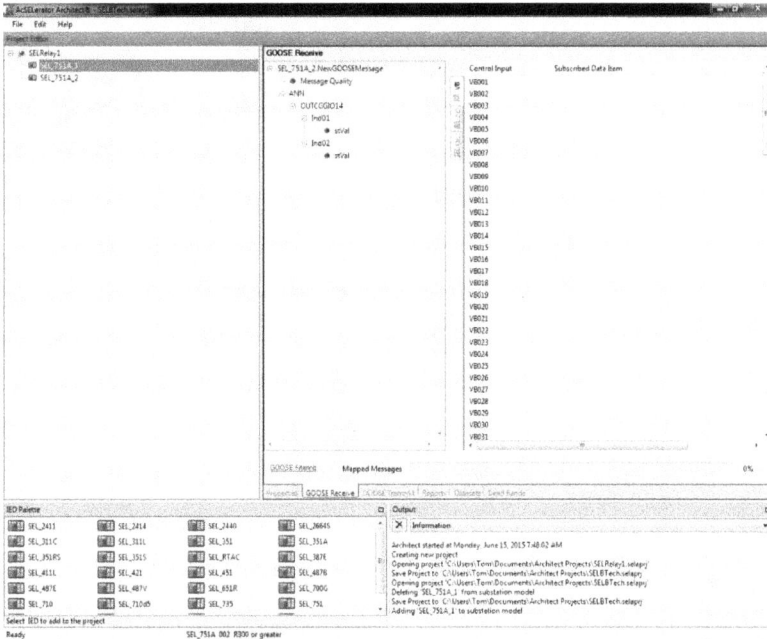

**Figure D.7: Subscription to GOOSE Message Data**

xvi) Click "Next", and your device is ready to receive GOOSE messages from "SEL_751A_2".

xvii) Ask the instructor to run the motor that is monitored by SEL relay SEL_751A_2, to test your configuration and programming.

xviii) Demonstrate your results to the instructor.

xix) Now program your IED to turn on output 1 on block C only if it receive a "1" from the GOOSE message and the power consumption on your side is less than P Watts.

**Notes:** Power consumption reading of your SEL Relay unit is varied by varying the light intensity of Lamp 1 using the light deeming switch on the unity. Power consumption tag is found in the following directory: Analog Quantities – Instantaneous Metering. Its value is represented by symbol "P". Note that this programming is done through AcSELerator QuickSet. For details refer to Section D.3.2.

[213]

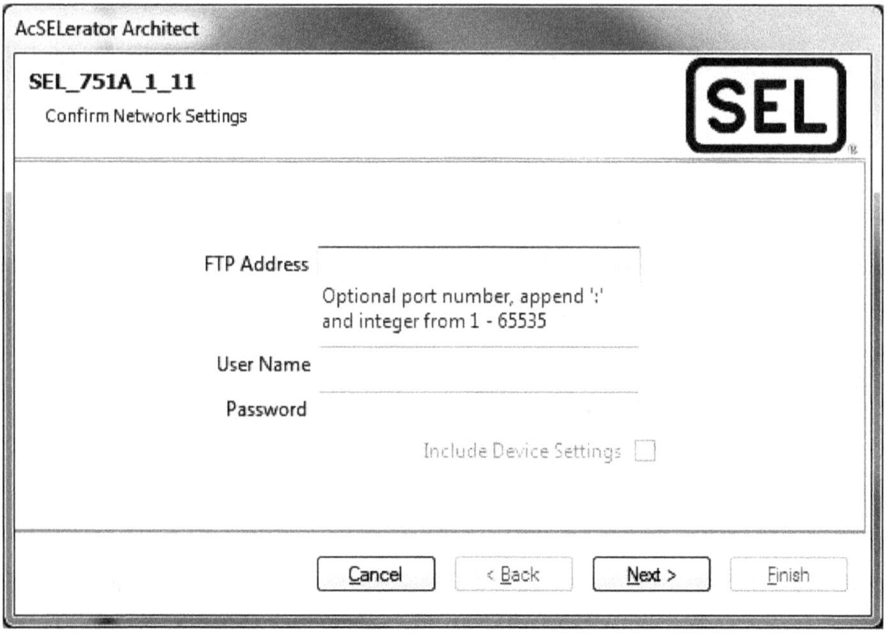

**Figure D.8: Transferring GOOSE Message Configuration to IED**

xx)   Set the value of P to 20 and ask the instructor to run the motor that is monitored by SEL relay SEL_751A_2, to test your configuration and programming

xxi)   Increase the value of P in steps of 20 and observe the behavior of lamp 2 with respect to the behavior of the motor and power consumption of the SEL relay unit

xxii) Explain your observations in D3.3xxi above.

# Appendix E

## Lab E: OPC Server Configuration Laboratory

### E.1    Objective

The main objective of this laboratory is to learn how to configure OPC servers to access data from various systems automation devices. To demonstrate the achievement of this objective, record the data (including screen shots) that shall help you to explain your observations and practice.

### E.2    Materials

The following materials are required for this laboratory:
i)      MicroLogix 1400 Temperature Control Unit with SEL 751 A Relay
ii)     Personal Computer ("the PC") with KEPServer OPC Server Software

### E.3    Procedure

### E.3.1    Data Access for Manufacturing Automation

This part of the lab is based on the MicroLogix 1400 Temperature Control Unit. You shall access this unit through the lab network as shown in Figure E.1. While the unit has many components, all one needs to access so as to control it, is the data or tags of the MicroLogix 1400 PLC. Those tags are provided in this laboratory.

### E.3.2    Configuration of KEPServer to Access Data from the MicroLogix 1400 PLC

i)      Locate and run KEPServer OPC Server application.

ii)     Click on "File", then on "New", if you get any warning about replacement of "Runtime" project, click on "No, Edit offline". If prompted to save, click "No". The Window should now look as shown in Figure E.2.

iii)    Click on "click to add a channel" as shown in Figure E.2. New window opens with channel name assigned to "Channel 1". You can change this name to one that has meaning with respect to your system (Figure E.3).

iv)     Click "next" on Figure E.3, and then click on the dropdown window to select "Allen-Bradley ControlLogix Ethernet".

## Remote Access: Temperature Control and Energy Monitoring System

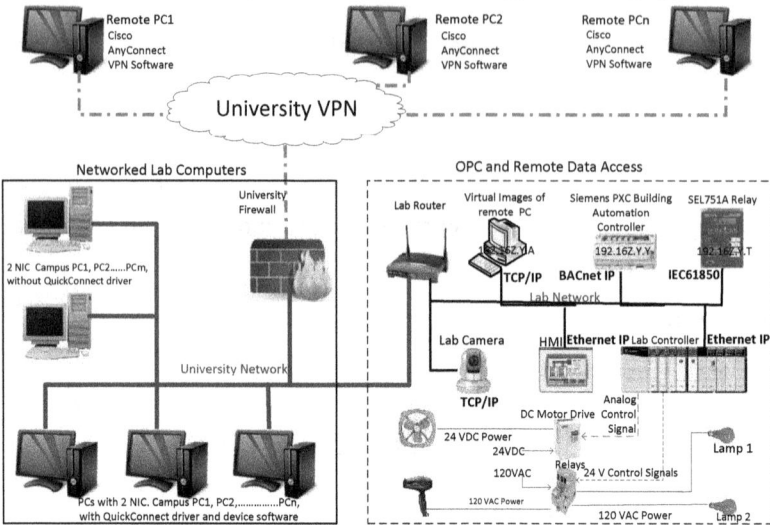

Figure E.1: IIoT lab Unit

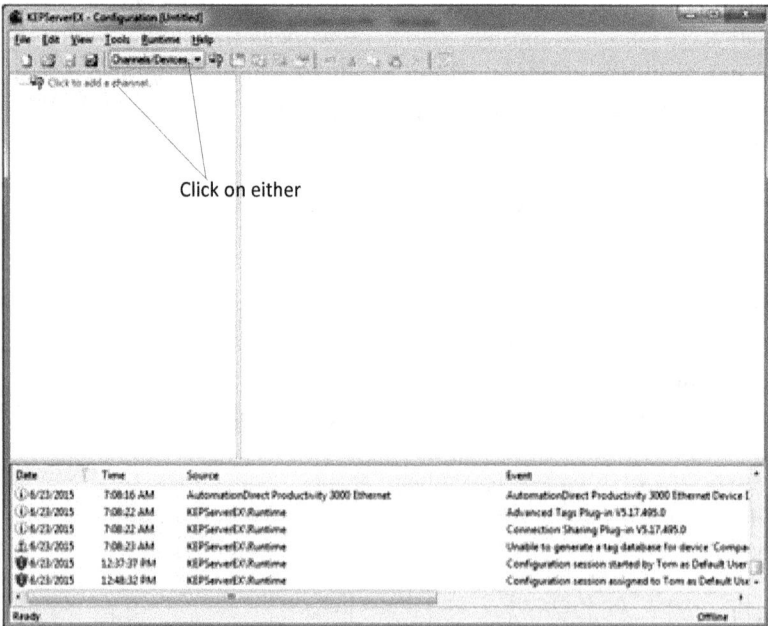

Figure E.3: KEPServer OPC Server Main Window

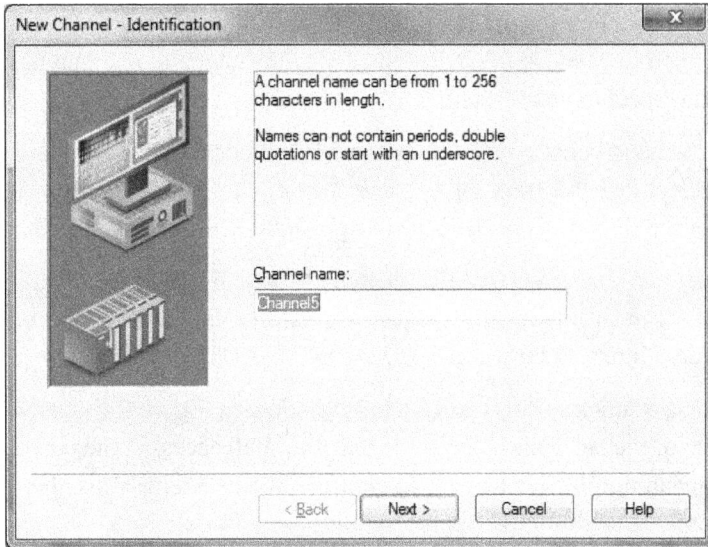

**Figure E.4: Channel Configuration Window**

v)     Leave the remaining settings at their default by clicking "next" until you click "Finish". The window now looks as shown in Figure E.4.

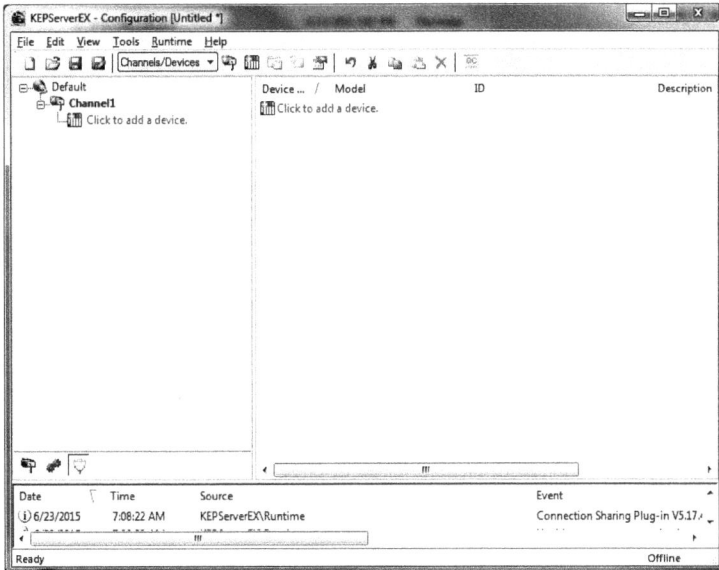

**Figure E.5: KEPServer OPC Server Communication Channel**

[217]

vi)   Click on "click to add a device". New window opens with device name assigned to "Device 1". You can change this name to one that has meaning with respect to your system.

vii)  Click "Next" on the new window, and then click on the dropdown window to select "MicroLogix 1400", and then click "Next". Under "Device ID" enter the IP address of the MicroLogix 1400 PLC as **130.113.130.93**.

viii) Leave the remaining settings at their default by clicking "next" until you click "Finish". Now the device is available in the project tree, and the window looks as shown in Figure E.5.

ix)   Click on "click to add a static tag". Window in Figure E.6 opens. Enter the name of the tag. This is the name that you shall access in the client. It should be meaningful with respect to your project. "Address" is the tag in the MicroLogix 1400 PLC.

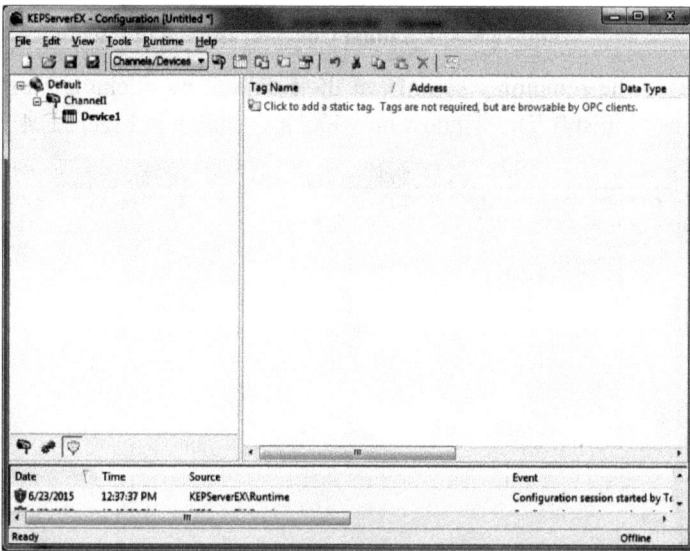

**Figure E.6: KEPServer OPC Server Device Configuration**

x)    In this lab you shall configure the following tags:

- Name: Temperature; Address: F8:3
- Name: PowerLimit; Address:  F8:17
- Name: PowerLimitSetting; Address:  F8:16

[218]

- Name: Lamp1Switch; Address: B3:0/0
- Name: Lamp1Status; Address: O:0/0
- Name: Lamp2Switch; Address: B3:0/1
- Name: Lamp2Status; Address: O:0/2
- Name: FanStatus; Address: B3:0/15
- Name: HeaterStatus; Address: O:0/2

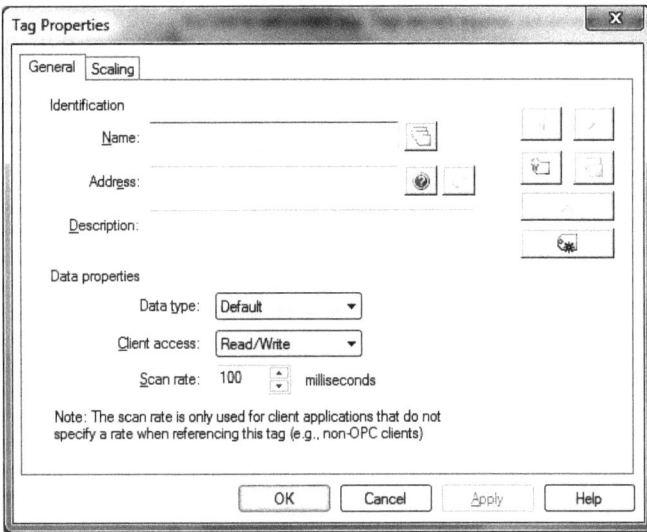

**Figure E.7: KEPServer OPC Server Tag Configuration Window**

xi) After entering the tag name and tag address, click "OK" and go to steps E.3.2 (ix) and E.3.2 (x) to enter the next tag, until when all tags are configured.

xii) After configuring the tags, the main window of the server looks as shown in Figure E.7.

xiii) Save the configuration on your memory stick by clicking on "File", and then "Save As".

xiv) KEPServer OPC server has an inbuilt "Quick" client that enables you to view the data associated with your tags.

xv)   Click on "Quick Client" (Figure E.7) to launch the quick the quick client. Then click on "Channel 1.Device 1" in the main project tree window of Figure E.8.

xvi)  To receive video and sound feed from the equipment you are controlling, open a web browser and enter the following IP address: **130.113.130.94**.

Ask the instructor to give the username and password.

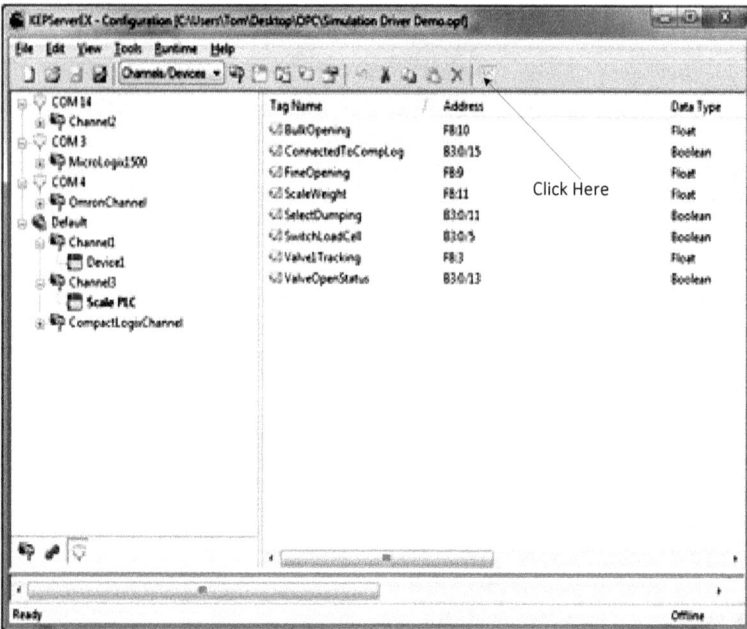

**Figure E.8: OPC Data Tags**

xvii)  Right click on Channel 1.Device 1. Lamp 1 Switch, and on the dropdown window click on "Synchronous Write…". The window in Figure E.9 opens. Double click under "Write Value", type in "1" press "Enter", if the lamp on the Micrologix 1400 Temperature Control Unit does not go on, click on apply. The lamp should go on. Repeat this step, but next time type in "0" to turn off the lamp. In both cases check the lamp status tags.

xviii)  Ask the instructor or technician to turn on the heating and cooling process, and monitor the temperature tag.

xix)  You now try controlling lamp 2 and monitoring the status of the lamps, fan, and heater.

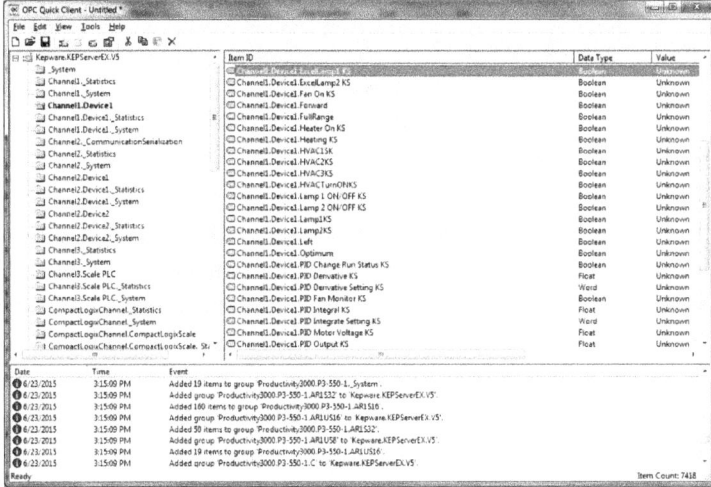

**Figure E.9: KEPServer Quick Client**

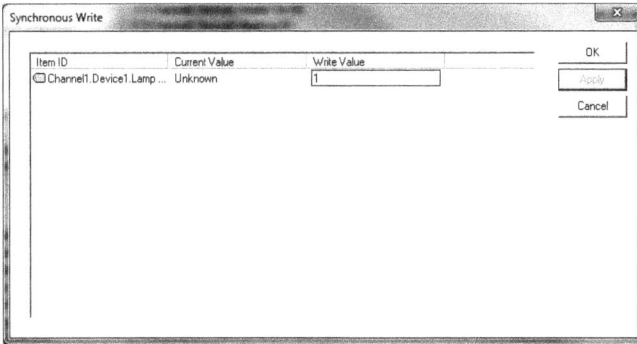

**Figure E.10: OPC Tag Writing**

xx)  Show the instructor your client, take its screen shot for your report and then close it.

xxi)  Save your server configuration again and close it. You shall need the configuration in the next lab.

### E.3.3    Configuration of IEC61850 MMS Client OPC Server

i)  Locate and run KEPServer OPC Server application.

ii) Click on "File", then on "open", the find your save KEPServer file from Section 3A.3.2 and open it.

iii) Click on "add a channel". New window opens with channel name assigned to "Channel 2". You can change this name to one that has meaning with respect to your system (Figure E.3).

iv) Click "next" on Figure E.3, and then click on the dropdown window to select "IEC61850 MMS Client".

v) Click "Next", open the dropdown menu under "Network Adapter" and select network card that has IP address of the form "130.113.130.X".

vi) Leave the remaining settings at their default by clicking "next" until you click "Finish".

vii) Click on "click to add a device" under "channel 2" in the project tree. New window opens with device name assigned to "Device 1". You can change this name to one that has meaning with respect to your system.

viii) Click "Next" on the new window until you see the parameters in Figure E.10.

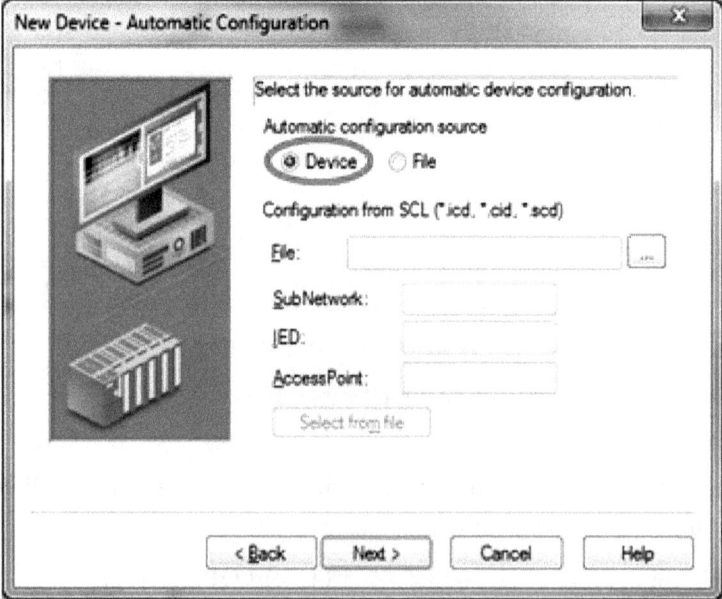

**Figure E.11: Configuration of IED Parameters**

ix)  Select "Device" as the "automatic configuration source".

x)  Click "Next", then enter the IP address of the IED which is **130.113.130.95** (Figure E.11).

xi)  Leave the remaining settings at their default by clicking "Next" until you click "Finish".

xii)  Click on "SAVE AS" to save the configuration. **YOU MUST SAVE THE CONFIGURATION NOW AS IT MAY NOT BE POSSIBLE TO POPULATE THE TAGS WITHOUT SAVING.**

xiii)  Close the configuration, and open it again by double clicking the saved file.

xiv)  To add tags, right click on the device name, **Device 1**, then click on "Properties". Window shown in Figure E.12 opens.

xv)  Click on "Database Creation", and then on "Auto Create". The tags shall be automatically populated. You notice that there are many tags. These tags were created in the device by default. AcSELerator Architect can be used to select the tags that are communicated through MMS protocol.

xvi)  Open the quick client, and then ask the instructor to run the motor system monitored by the SEL 751A relay.

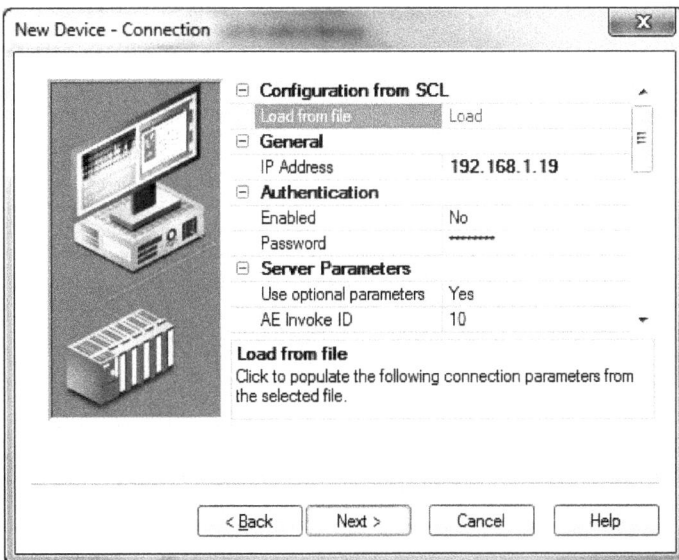

**Figure E.12: IED IP Address**

[223]

xvii) Browse the data associated with the relay tags. Identify the power, phase voltage, current, and frequency tags

xviii) Show the instructor your client, take its screen shot for your report and then close it.

xix) Save you server configuration again and close it. You shall need the configuration in the next lab.

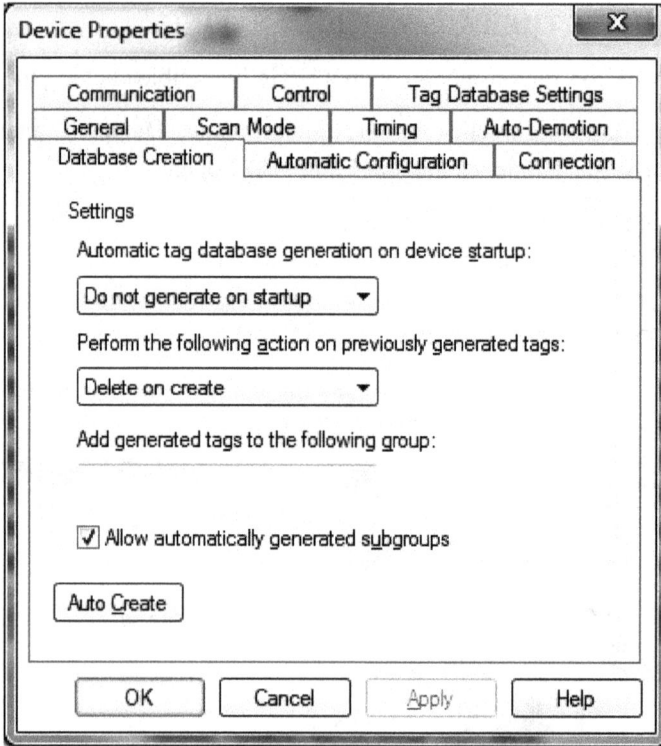

**Figure E.13: Auto Creation of Tags Database**

# Appendix F

## Lab F – OPC Data Access

### F.1      Objective

The purpose of this experiment is to learn how to integrate systems using OPC technology. In this laboratory, OPC DataHub is the OPC client, while KEPServer is the OPC server. To demonstrate the achievement of this objective, record the data (including screen shots) that shall help you to explain your observations and practice.

### F.2      Materials

The following materials are required for this laboratory:

i)      MicroLogix 1400 Temperature Control Unit with SEL 751 A Relay

ii)      Personal Computer ("the PC") with KEPServer OPC Server Software, as well as OPC DataHub Software

### F.3      Procedure

### F.3.1      Add OPC Server to OPC DataHub Client

i)      Locate and run OPC DataHub application by double clicking on its icon. You get a warning that after one hour the application shall shutdown because it has an evaluation license. Click "OK" on the warning.

ii)      The icon of the application now appears in the low toolbar (Figure F.1). Double click on the icon to open the main window of OPC DataHub shown in Figure 8.9.

iii)      Click on "OPC" on the DataHub main window, then click "Add". Window in Figure 8.10 opens. Set parameters as shown in Figure 8.10, and then add KEPServer Ver 5.1 OPC server as explained in Section 8.9.3.

iv)      Click "ok' on window in Figure 8.10 and then click "ok' on window in Figure 8.9. Both windows close. Reopen main window by click on the icon in the toolbar.

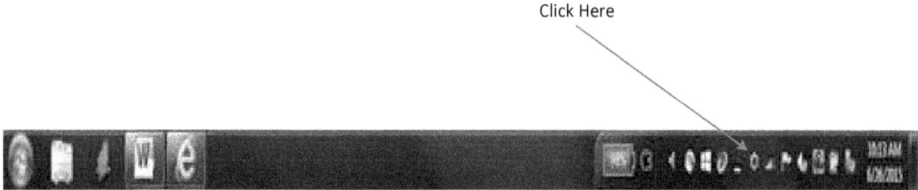

Click Here

**Figure F.1: OPC DataHub Icon in Toolbar**

## F.3.2    View Data

i)      Once the OPC server is running data is being transmitted to the OPC
        DataHub client.

ii)     To look at this data Click the "General" button to go back to the General
        Settings Screen  Figure F.2)

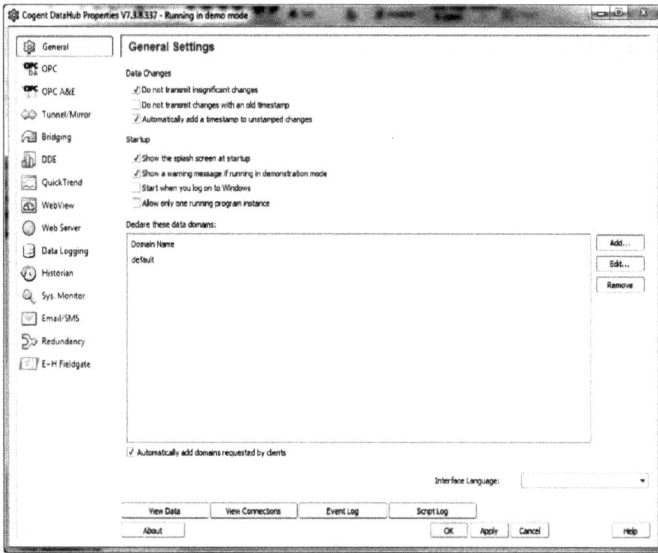

**Figure F.2: General Settings Window**

iii)    On the General Settings screen there is a section labeled "Data Domain
        Name" (the "default" domain should be there by default). You can add
        any other domains that you could have created by clicking "Add".

[226]

iv)  Below the data domain section click on "View Data". The Data Browser window in Figure F.3 opens.

v)  The data on the window should be updating. Show the instructor your results.

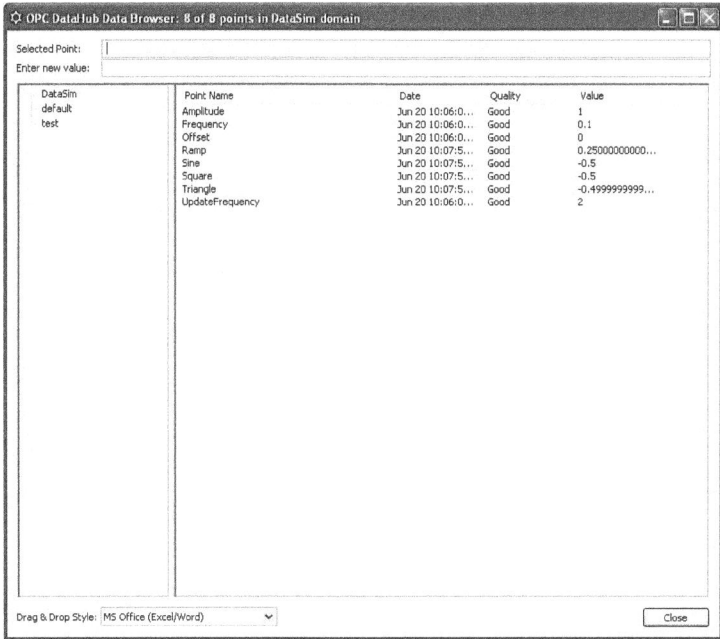

**Figure F.3: Data View Window**

vi)  On the Data Browser window click on points (variables), if write capabilities are configured the variables values, can be changed through the Data Browser window.

vii)  To do this simply click on the point and enter a new valid value in the "Enter new value box" at the top of the screen and press "Enter".

## F.3.3    Data Bridge Procedure

Carry out the following steps to bridge the system power tag "METMMXU1$MX$TotW$instMag$f" of the SEL 751 relay to the "PowerLimit" tag on the MicroLogix PLC (the power limit tag turns off lights is the power consumption of the automation system goes above a value set by the "PowerLimitSetting" of the PLC):

[227]

i)   Click on "Bridging" in the main menu of OPC DataHub, and click on "Configure Bridges". Window similar to one shown in Figure F.4 opens.

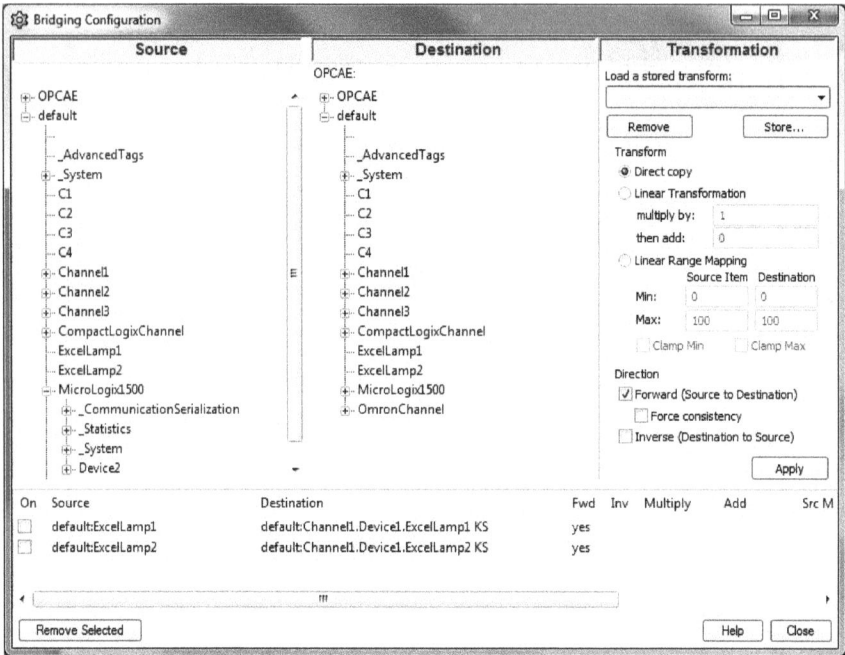

**Figure F.4: OPC DataHub Bridging**

ii)  Figure F.4 shows the source of data (the data you want to bridge) and the destination (where you want the bridged data to go). You notice that you have the same data in the source and in the destination. This means that any data can be bridged. Open the data groups by clicking on the "+" sign in front of the data group name.

iii) Click on the data point you want to bridge on the source side, then click the data point you want to bridge data to on the destination side.

iv)  Click on "Apply" and a new bridge is formed, and you can see it in the bottom window of Figure F.4.

OPC DataHub supports transformation of bridge data. You notice from the transformation window of Figure F.4 that the data bridging in the above lab steps

[228]

lead to "direct copy" of the power reading from the SEL relay to the power limit tag of the PLC. Moreover, OPC DataHub can modify data using linear transformations of the form Y=mX+C, or scale it using linear range mapping.

**F.3.4    Data Bridge Testing**

i)    Carryout steps described in Section E.3.2 to set the value of the tag "PowerLimitSetting" to 100W.

ii)   Ask the instructor to turn on the lights and then the automated heating/cooling system.

iii)  Monitor the power tag, and what happen to the lights when the power consumption exceeds 100W.

iv)   Increase the value of the "PowerLimitSetting" in steps of 50W, and monitor the power and the lights.

v)    Keep screen shots that hall help you to describe and discuss your observation in the lab report.

**F.3.5    OPC DataHub HMI Procedure**

i)    Click on "WebView" on the OPC DataHub main window (Figure 8.9). Window in Figure F.5 opens. Click "Refresh" and the check all the boxes besides the data domain so as to include all the domains in the web based HMI.

ii)   Click "Apply".

iii)  Click on "Launch WebView in a browser" and wait for the browser to open. If it does not, open a web browser and type in http://localhost/Silverlight/DataHubWebView.asp. Window in Figure F.6 opens.

iv)   The username is admin and the password is also admin. Type in the username and password and click "connect".

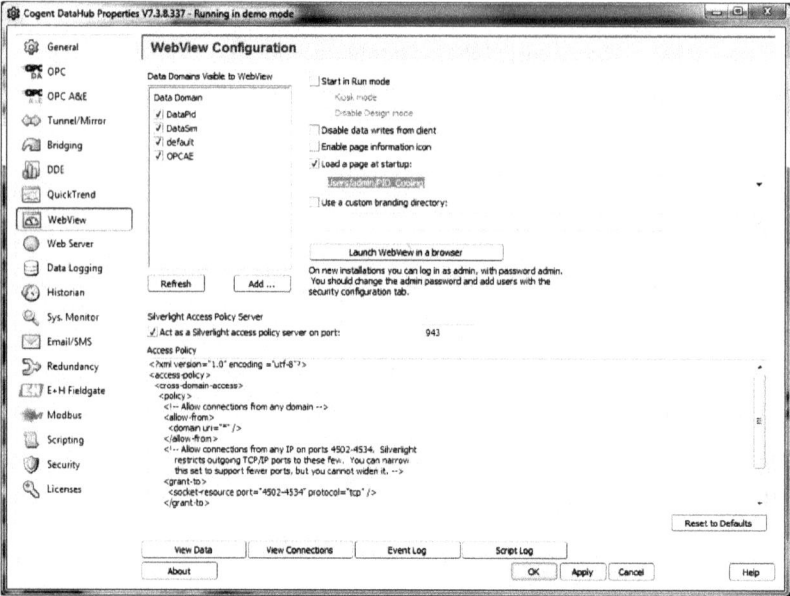

Figure F.5: WebView Configuration

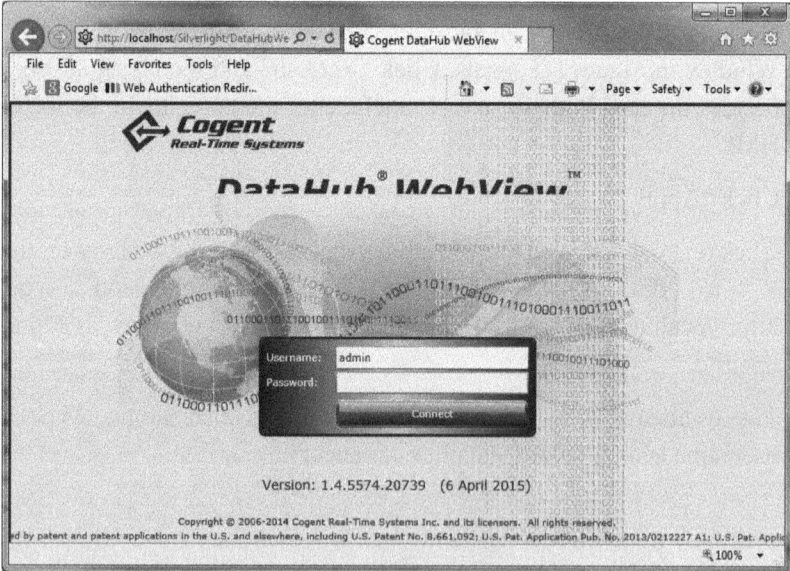

Figure F.6: Login Window of DataHub WebView

v)    Click on "File" and double click on "New". Window in Figure F.7 opens.

[230]

vi)   Click on "Toggle Button" marked 1, then click on "Basic Properties", marked 2 in Figure F.7.

vii)  Open the value binding by clicking on "open" icon marked 3 in Figure F.7, then click on the "open" icon next to the "Binding" and click on "Point". Binding window opens.

viii) Type in the name of the tag you want to bind to the "Toggle Button". In this case type in "Lamp1Switch". This is the name you signed the tag that controls Lamp1. Note that after typing in two first letters of the tag name, all tags with the letters in tag name open in a dropdown window. Click on the one you want to bind to the symbol. Tags appear with their folder path properly defined.

ix)   Click on "Shining Light" marked 7 in Figure F.7 and follow steps vii to viii to bind the light to the "Lamp1Status" tag.

x)    Click on "Text Entry Field" marked 6 in Figure F.7 and follow steps vii to viii to bind the text entry field to the "PowerLimitSetting" tag.

xi)   Click on "Text Entry Field" marked 6 in Figure F.7 and follow steps vii to viii to bind the text entry field to Phase A power tag of the SEL relay tag.

xii)  Notify the instructor that you are ready to test your HMI, then click on the "Run" icon marked 8 in Figure F.7.

**F.3.6   OPC DataHub HMI Testing**

i)    Set the value of the tag "PowerLimitSetting" to 100W by typing it into the setup text entry field on the HMI.

ii)   Notify the instructor that you are ready to test your HMI. Turn on the lights and ask the instructor to turn on the automated heating/cooling system.

iii)  Monitor the power tag in the power field on the HMI, and what happen to the lights when the power consumption exceeds 100W.

iv)   Increase the value of the "PowerLimitSetting" in steps of 50W, and monitor the power and the lights.

v)    Keep screen shots that hall help you to describe and discuss your observation in the lab report.

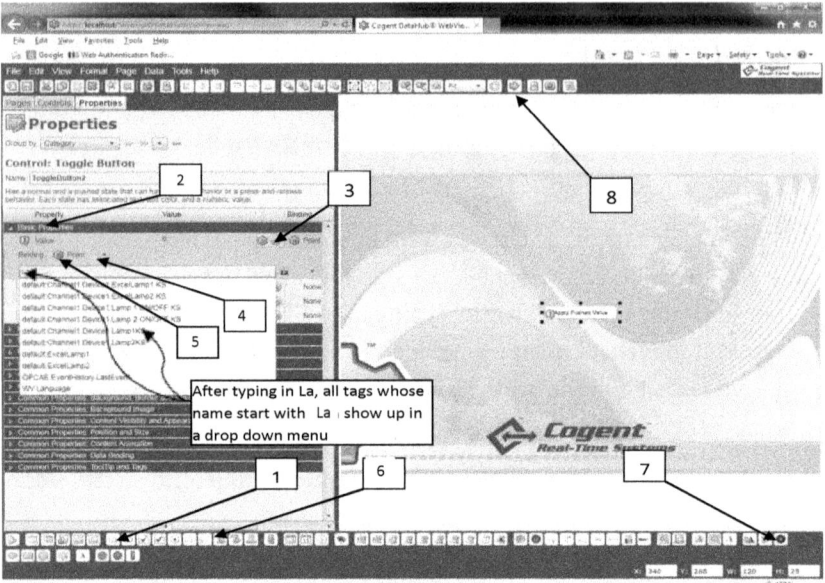

After typing in La, all tags whose name start with La show up in a drop down menu

**Figure F.7: Configuration of WebView Items**

# Appendix G

## Lab G – Configuration of BACnet IP Devices

### G.1     Objective

The purpose of the laboratory is to learn how to configure the communication settings of BACnet IP devices.

### G.2     Materials

The following materials are required for this laboratory:

i)     PXC16 Compact Building Automation Controller
ii)     Ethernet Cable
iii)     24VAC Power Adapter
iv)     RS-232 to RJ11 Cable
v)     Tera Term HyperTerminal
vi)     Innea BACnet Explorer

### G.3   Procedure

i)     Wire 24VAC adapter to 24V~ INPUT and GND of the PXC controller. Ensure jumper wire between neutral and GND is present. Check status of unit when powered, RUN should be solid green and no faults present. Replace AA battery backup if needed. To replace battery, press down lever on front panel of device and remove cover. Note polarity of AA battery and replace if required. Note that if the controller is part of a laboratory prebuilt unit, this step may not be required.

ii)     Connect the supplied cable to the HMI port on the unit and run the other side to the RS-232 cable coming from the PC's serial COM port. Check with device manager for correct COM. Generally, COM1 is set aside for this serial communication, but it is a good habit to double check. Make a note of the COM number.

iii)     To add the Ethernet route for the device, connect a supplied Ethernet cable to the Ethernet port on the PXC at the top right corner. Run this cable to the switch at your station which then runs to your LAN card on the PC at your station. Note that if the controller is part of a laboratory prebuilt unit, you may just need to connect the PC Ethernet port to the unit's Ethernet switch.

[233]

iv)     Open Tera Term and click the serial radio button, use the corresponding COM noted in the previous step. Set up the configuration parameters for the communications as follows: 9600 Baud rate, 8 bits, no parity, 1 stop bit, and no flow control (see Figure G.1). Double check these settings under Setup -> Serial port... The previously mentioned configuration parameters should be default.

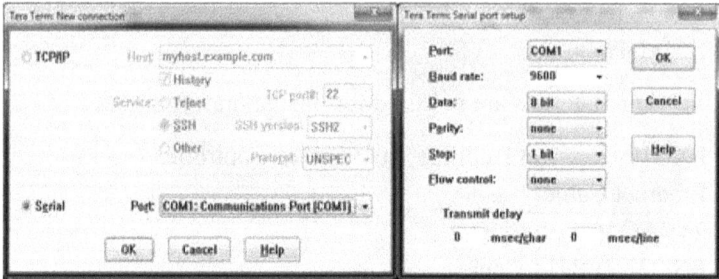

**Figure G.1: COM Configuration**

v)      Hit the enter key (or any key) in the terminal and the unit will respond with some options. To move through the menu in Figure G.2, use the capitalized letter in each command that is separated by a comma.

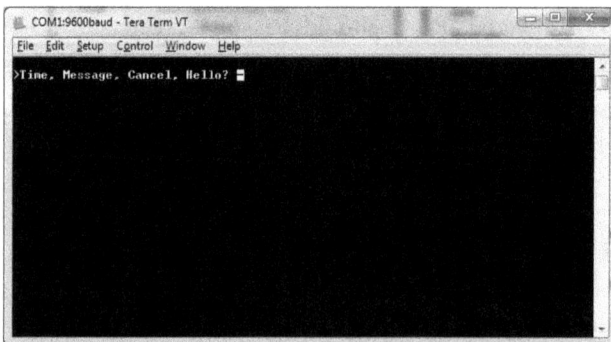

**Figure G.2: PXC Tera Terminal Main Menu**

vi)     First, hit H for Hello. This prompts user login followed by password. Use 'high' for user and pass to gain access to the device at the highest control level (low and med are alternative user accounts with less permissions) (see Figure G.3).

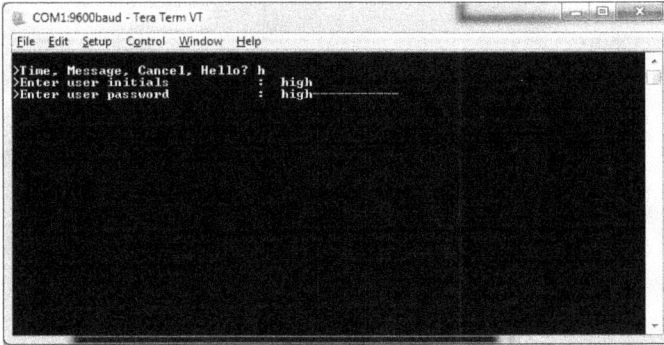

**Figure G.3: PXC Login Menu**

vii) To configure the Ethernet port of the controller, go to System (S) followed by Hardware (H) then Ethernet (E) and ipSettings (S); by pressing the letters in brackets (see figure G.4).

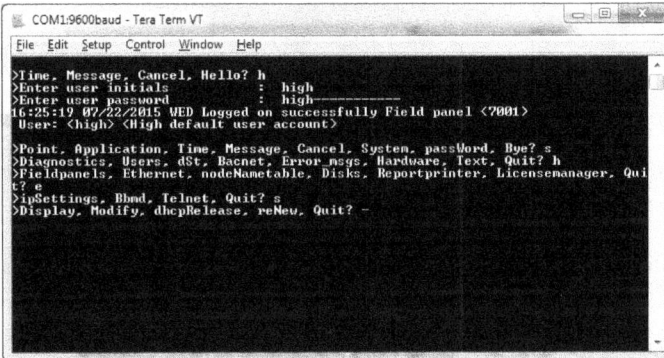

**Figure G.4: PXC Configuration Menu**

viii) Press D to display current connection settings. Note down the current IP address already configured and compare to desired network settings shown in Figure G.5.

ix) To set up a different IP for the unit go back by inputting Quit (Q) and going back to ipSettings (S) and then enter the Modify (M) command. Carefully go through each prompt according to desired settings. In this case, leave everything alone except for the IP address, hit enter to get through each of the preconfigured settings. Once you get through the first couple network settings, configure BACnet will come up. If you are happy with the previous display of BACnet settings (Instance number is

[235]

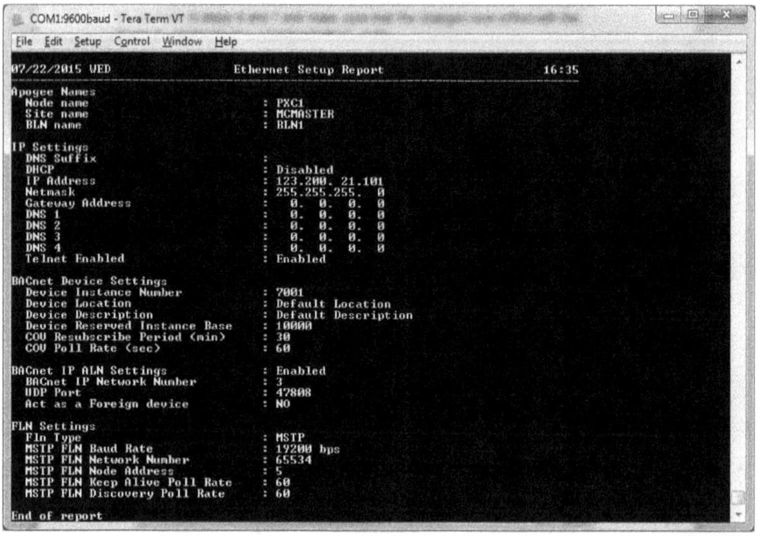

**Figure G.5: Network Connection Settings**

the main one to note) then hit N and the modification command
successfully changed the ipSettings. If prompted for a Cold Start to reset
the device, hit Yes (Y). Now double check that the settings were changed
for the Ethernet Setup (see Figure G.6).

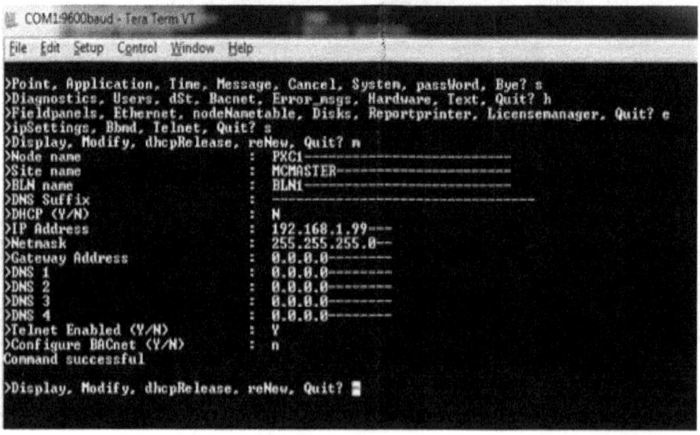

**Figure G.6: Window for Changing Ethernet Communication Settings**

x)    Go back to step viii and make sure that the changes took effect with the
device by displaying the current ipSettings again (see Figure G.5).

[236]

xi) Now that the device has been configured to be on the same network as the PC LAN, its information can be explored through BACnet explorer application. Launch the Innea BACnet Explorer and click the Explore network button on the top toolbar. The PXC should then be displayed on the network; ID, name, IP, and port are all displayed right away (see Figure D.7).

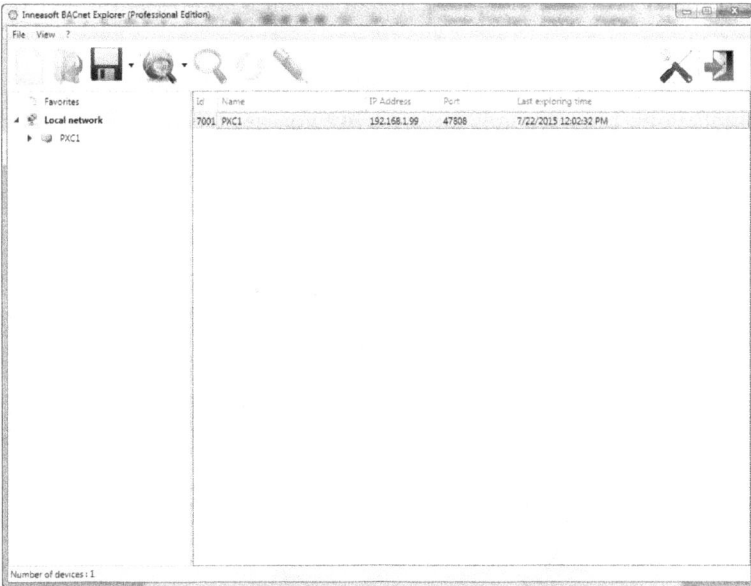

**Figure G.7: Innea BACnet Explorer Window Showing PXC Configuration**

xii) Expand the PXC tree on the left under Local Network to gain more information about the device. Go to the device tree and click the PXC to see all of the properties associated with this particular device.

xiii) Since the device is now ready to be monitored, Test point can be set up, and be controlled through Innea BACnet Explorer. To do so, go back to Tera Term HyperTerminal and use (#) Shift-3 to get back to the main menu. If your login timed out, login again. Then go to Point (P), Edit (E), and then Add (A).

xiv) This is where the properties for the point database get populated. First is the Point System Name (this is how you reference your point similar to tag name in standard PLC), then comes instance number. This is a unique identifier for each device. Just hit enter and the instance number will default to 0 or

[237]

the lowest value available. In the case of the blank spaces, just hit enter key as they will default to correct values. The main properties here are point system name, point name, type, field panel, FLN and Point. FLN = 0 for direct local network control from this device, Point is the number associated with the I/O you are trying to use and Field Panel is the instance number for the Device. Point numbers can be found directly on the PXC where you have I/O points 1-13 along the bottom and Digital Output 14, 15, and 16 on the top. These top 3 points are relays. Point system name and point name can be identical, system name is the main identifier we use when referring to the point. Point type depends on what I/O you are trying to use (Figure G.8). Refer to APOGEE Field Panel User Manual pages 253 – 260 for relevant point setup information.

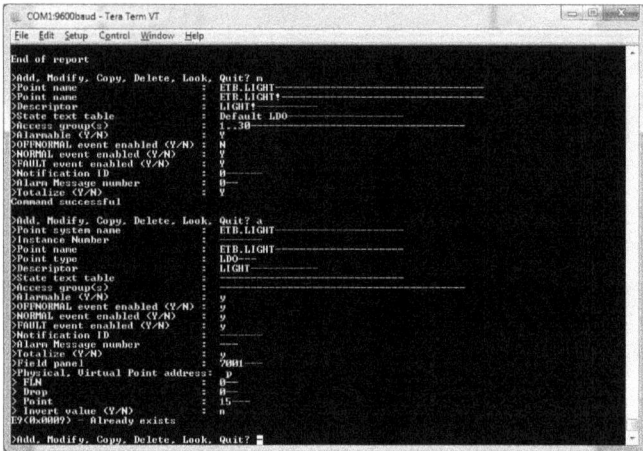

**Figure G.8: Configuration of PXC Point IOs**

xv)  Now that the point has been successfully added to our device, we can monitor and command the point from either the terminal window or with the BACnet explorer software. First, let's check its status with the terminal. Quit (Q) out of the Point editing and go to Log (L). Enter a (*) when prompted for Point Name, this asterisk is the command for requesting ALL points of any name. The window in Figure G.9 shows the state of the points, name, description, status and priority.

[238]

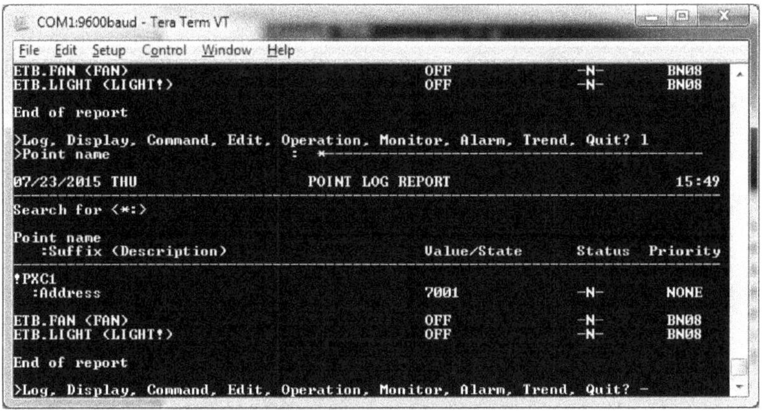

**Figure G.9: Point IO Status Information**

xvi) To command a point go to Command (C) and then Value (V) and input the Point name (Point system name). For the example in Figure G.9, The point IOs are ETB.FAN and ETB.LIGHT. Search for the point you named with the exact string; note that the HyperTerminal is not case sensitive. Enter ON in the new state and use priority level 8 for the highest priority available (see Figure G.10). (This priority is used to avoid other software from overriding the entry if a lower priority is used. Note that the higher priority identification number, the lower the priority - an inverse relationship.)

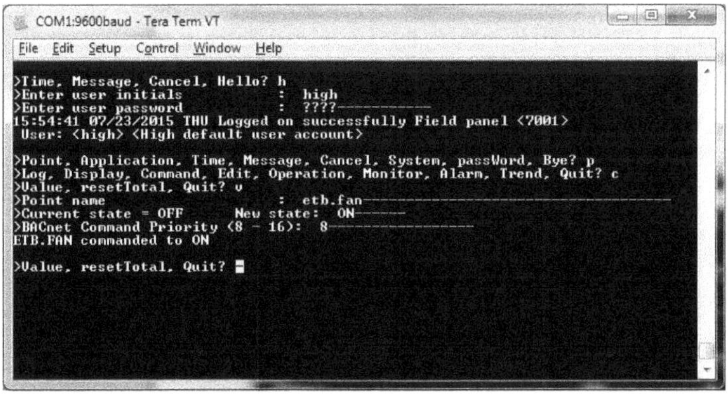

**Figure G.10: Window for Changing the State of Points**

xvii) To verify the effect of the command, check the physical output of the PXC controller to see if it has been activated and turned on the hardware. The command effect can also be verified by going to the point log. Go back to

[239]

Log (L) and then input either an asterisk for all point names or search for your specific point name. Change of status should be noticed in the log (see Figure G.11).

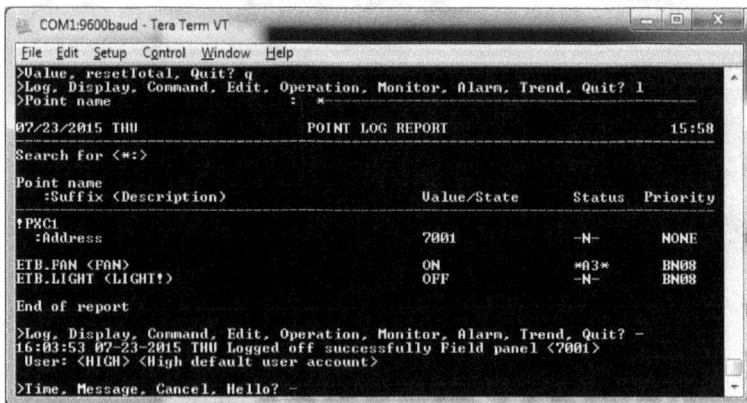

**Figure G.11: Change of Point Status**

*Note: The point will display alarms or faults in the log depending on how you initially configured the point when you added it. There are a few alarm prompts when adding a new point that activate when the point receives power. This is just a safety precaution so that operators acknowledge changes to the system in place.*

xviii) You can verify point command was successful with Innea BACnet Explorer as well. Go to PXC16 under the Local Network tree again and look under Binary Output (For the digital outputs) for your points. Click on the point you want to monitor and wait for the properties to load. The present value property is what should be monitored.

xix) The device can now be controlled and monitored through the terminal or monitored through the BACnet explorer. This concludes the initial configuration and familiarization with BACnet and PXC Compact functionalities.

*RELEVANT KEYSTROKEs: SHIFT-3 (#) is cancel!, CTRL-F moves cursor to right in terminal.*

# Appendix H

## Lab H: Programming Building Automation PXC16 Controller

### H.1 Objective

The main objective of this laboratory is to introduce students to the programming of building automation controllers using the PPCL language.

### H.2 Materials

The following materials are required to carry out this laboratory:

i) PXC16 Compact Building Automation Controller.
ii) Ethernet Cable.
iii) 24VAC Power Adapter.
iv) RS-232 to RJ11 Cable.
v) Tera Term HyperTerminal.
vi) Innea BACnet Explorer.
vii) 120VAC bulb and 24VDC relay.
viii) 120 VAC fan, 24 VDC relay, and a RTD temperature sensor.

### H.3 Procedure

i) Wire the lamp and the fan to the controller digital outputs through external 24VDC relays.
ii) Wire the RTD to the controller inputs.
iii) Set up the controllers as explained in steps i to vi in laboratory 4A.
iv) Configure the RTD input parameters as shown in Figure H.1, by following steps xiii and xiv in lab 4A.
v) To begin programming go to Application (A) and then PPCL (P). Once inside the PPCL menu options, go to Edit (E) and enter a program name of choice. In this case, 'program' was used. Followed by the field panel instance number such as 7001, and a priority of 8 (see Figure H.2).
vi) To check on the program in the controller (editing a previous program) go to Look (L) and enter number 1 for the first line of the program. Any other number less than or equal to the program can be entered to view the associated program line. If there is no program in the controller it responds with "E3 (0x0003) - not found".

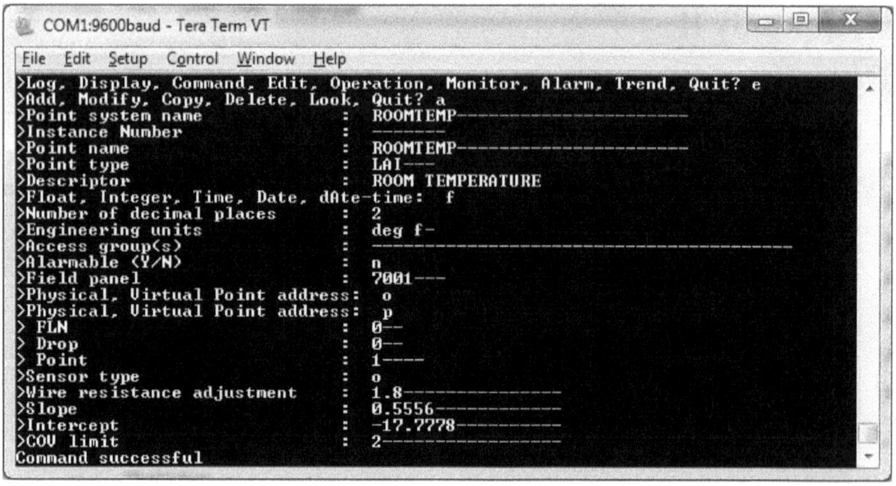

**Figure H.1: Configuration of the RTD Input**

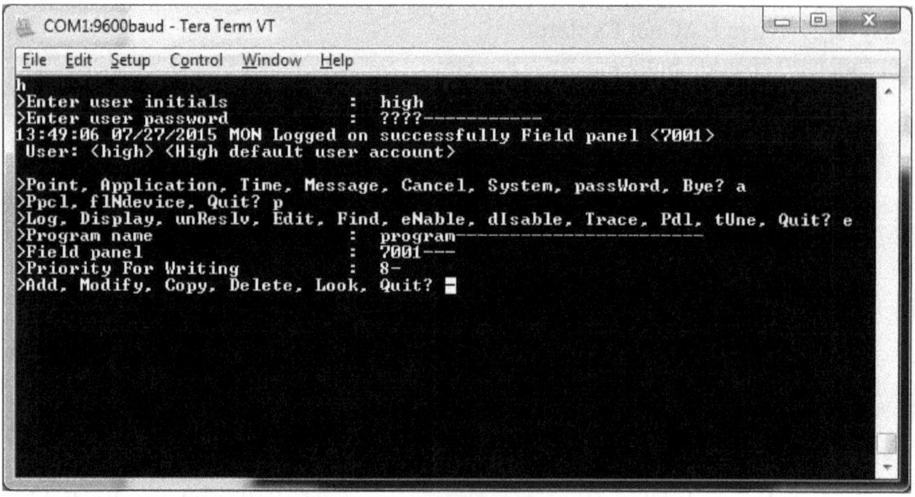

**Figure H.2: Program Configuration Window**

vii)  To add a line, press A for "add". This example starts with a comment. Therefore, type "1 C My first PPCL program" as your first line and press enter. Number "1" means that "My first program" is the first line of the program, and letter "C" means that My first program is a comment. Once the line has been compiled and checked for errors the terminal responds with "PPCL Line Added Program name: '*your program name*'" (see Figure H.3).

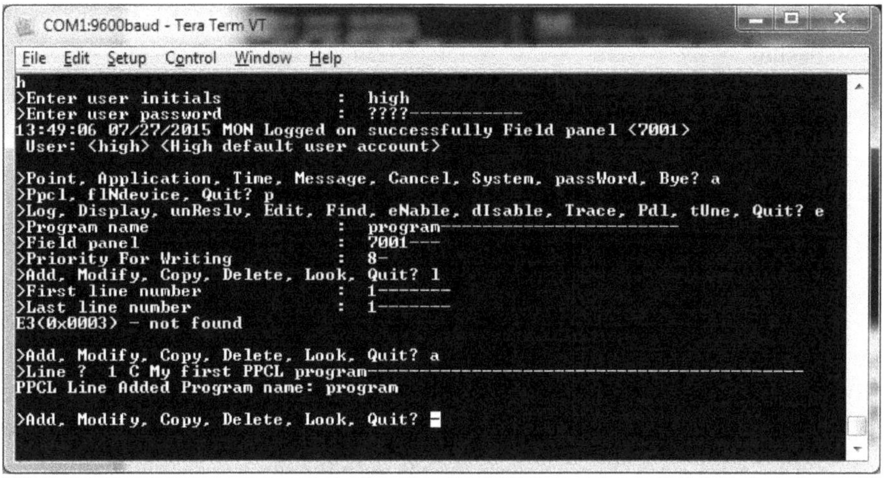

**Figure H.3: First PPCL Program Line**

viii) Note that without the number 1 or letter C, in the program, the compiler would have returned syntax errors.

ix) Now to double check that the line was added properly, go to Look (L) and enter first line: 1 and last line: 1. This returns the first line of our program as shown in Figure H.4.

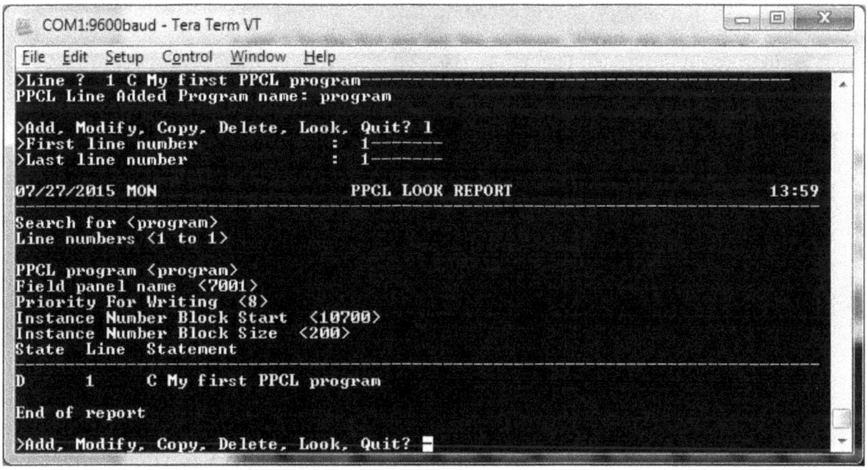

**Figure H.4: Viewing PPCL Program Line**

x) Every PPCL line has to be initiated with a line number, and the maximum length of lines is 67 characters including spaces and line numbering. Tera

Term shows the maximum characters allowed for each terminal option with dashes.

xi) Control logic can now be added to the program using predefined points: ETB.FAN – Digital Output Relay at point 14, ETB.LIGHT – Digital Output Relay at point 15 and ROOMTEMP – Analog Input at point 1. Refer to lab 4A for information on how to configure controller points.

xii) Choose a temperature setpoint, for example 70 degrees Fahrenheit. When temperature goes above this value, turn on the fan. If temperature reading is less than 70 Fahrenheit, turn on the light (simulating a heater). Manipulate the temperature reading by applying body heat to the RTD.

xiii) Add (A) the following new line to the program as shown in Figure: H.5: "2 IF(ROOMTEMP.GT.70) THEN ON("ETB.FAN") ELSE OFF("ETB.FAN")". This corresponds with the fan control and will turn on the fan if temperature reading is greater than 70. The else command turns off the fan if temperature is less than 70. This else clause is necessary in order to have desired control of the fan.

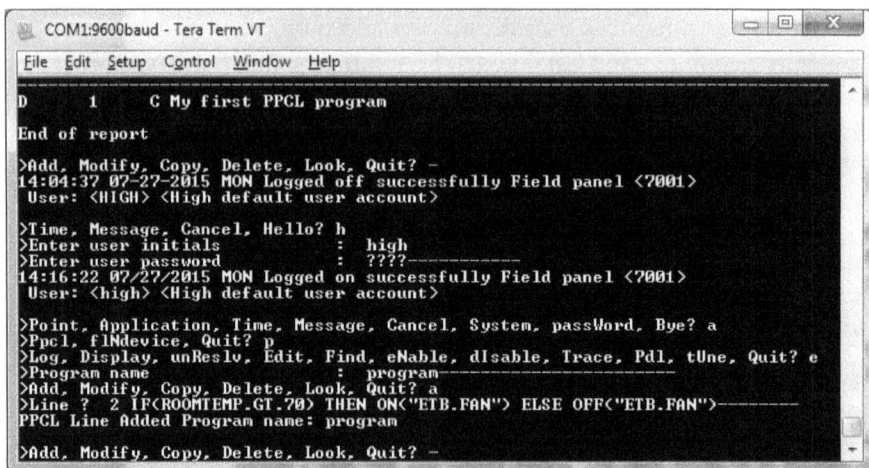

**Figure H.5: Fan Control Logic**

xiv) To add the heating control logic to the program, select Add (A) then type "3 IF(ROOMTEMP.LT.70) THEN ON("ETB.LIGHT") ELSE OFF("ETB.LIGHT")" as shown in Figure H.6.

[244]

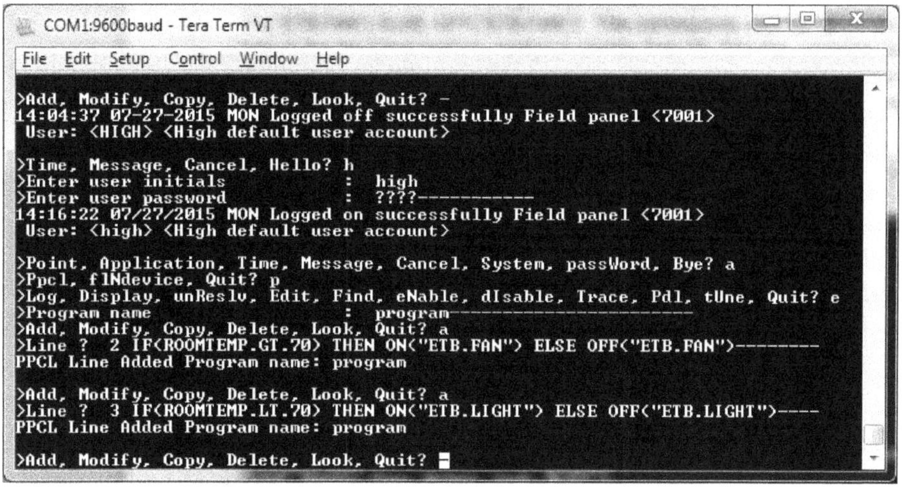

**Figure H.6: Heater (Lamp) Control Logic**

xv)  Double check that the new line additions went through by using Look (L) and entering again and enter first line: 1 and last line: 3. This returns all 3 lines of PPCL code (see Figure H.7).

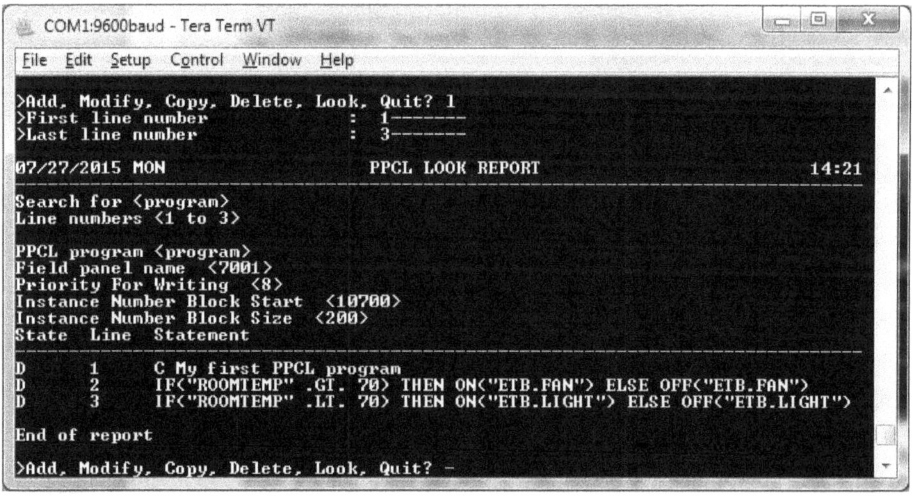

**Figure H.7: Viewing PPCL Code**

xvi)  Now the program needs to be enabled (to run). Back out of the program editing to the PPCL menu options by selecting quit. From the PPCL menu options choose eNable (N) and enter the program name and line numbers to

[245]

enable. Once enabled successfully the terminal responds with: "PPCL lines enabled for program *'your program name'*". Initially, the light should be turned on to simulate the heating process as the temperature would be below the setpoint. Wrap your hand around the temperature probe (RTD) to raise the temperature. The light should turn off, and the fan should turn on once the temperature goes above the setpoint.

xvii) *RELEVANT KEYSTROKES & BACKGROUND: SHIFT-3 (#) is cancel!, CTRL-F moves cursor to right in terminal. Refer to APOGEE Field Panel User's Manual for point setup and example programs; and to Siemens APOGEE PPCL User's Manual for PPCL syntax help.*

xviii) To display the definition of the points (tags), go to Point (P), Display (D), Definition (D), Name (N). Enter the point name and field panel when prompted. The following fields should be displayed, double check that the values are as desired! The Condition value should be registering as –N– which indicates a normal state. Instance number is what is used to address points in BACnet programs. (For this example, BAC_7001_AI_0 would refer to the temperature sensor connected at the analog input with the first default instance number of 0) (see Figure H.8).

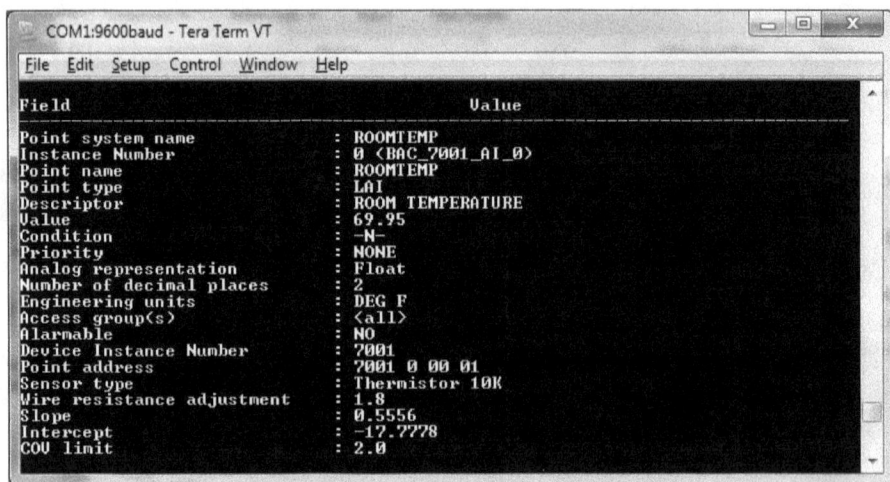

```
COM1:9600baud - Tera Term VT
File  Edit  Setup  Control  Window  Help

Field                               Value

Point system name                 : ROOMTEMP
Instance Number                   : 0  (BAC_7001_AI_0)
Point name                        : ROOMTEMP
Point type                        : LAI
Descriptor                        : ROOM TEMPERATURE
Value                             : 69.95
Condition                         : -N-
Priority                          : NONE
Analog representation             : Float
Number of decimal places          : 2
Engineering units                 : DEG F
Access group(s)                   : (all)
Alarmable                         : NO
Device Instance Number            : 7001
Point address                     : 7001 0 00 01
Sensor type                       : Thermistor 10K
Wire resistance adjustment        : 1.8
Slope                             : 0.5556
Intercept                         : -17.7778
COV limit                         : 2.0
```

**Figure H.8: Display of Point (Tag) Data**

**Lab H: PART 2- Configuration of KEPServer Data Access for PXC16 Controller:**

**H.4    Objective**

The main objective of this part of the laboratory is to learn how to configure OPC data access for the PXC16 building automation controller using BACnet IP communication protocol. Such data access is required in the development of building monitoring and SCADA systems. To demonstrate the achievement of the laboratory objective, record the data (including screen shots) that shall help you to explain your observations and practice.

**H.5    Materials**

The following materials are required for this laboratory:
iii)    PXC Controller.
iv)    Personal Computer ("the PC") with KEPServer OPC Server Software

**H.6    Procedure**

i)    Locate and run KEPServer OPC Server application.

ii)    Click on "File", then on "New", if you get any warning about replacement of "Runtime" project, click on "No, Edit offline". If prompted to save, click "No". The window should now look as shown in Figure H.9.

iii)    Click on "click to add a channel" in Figure H.9. New window opens with channel name assigned to "Channel 1". You can change this name to one that has meaning with respect to your system (see Figure H.10).

iv)    Click "next", and then click on the dropdown window to select "BACnet/IP".

v)    Leave the remaining settings at their default by clicking "next" until you click"Finish". The window in Figure H.11 opens.

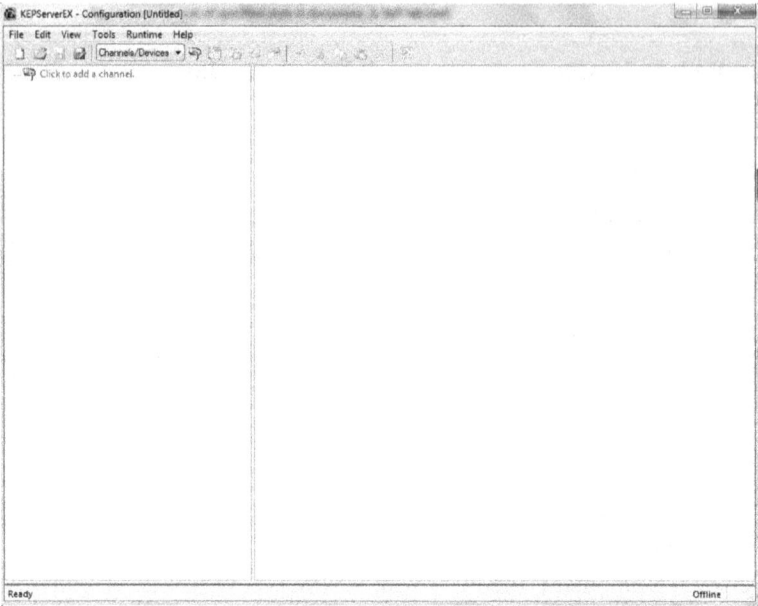

**Figure H.9: New KEPServer File Window**

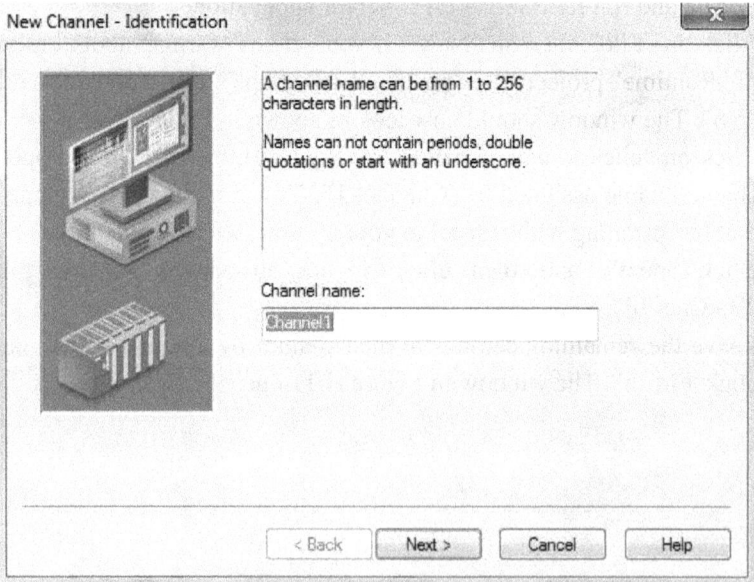

**Figure H.10: New KEPServer Communication Channel**

vi)    Click on "click to add a device". New window in Figure H.12 opens with device name assigned to "Device 1". You can change this name to one that has meaning with respect to your system.

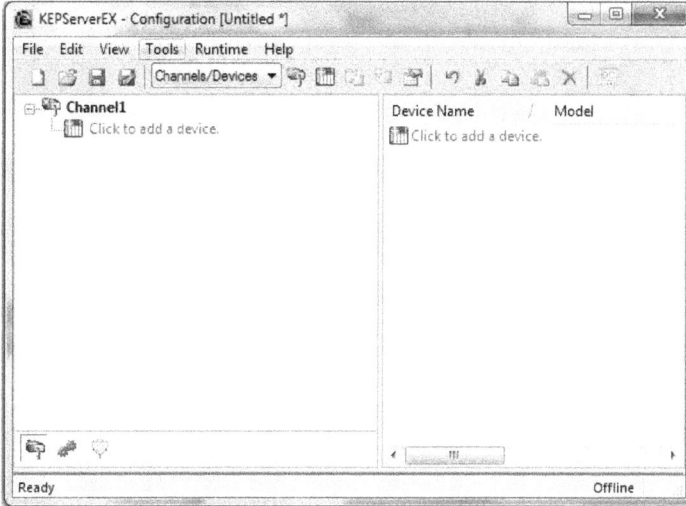

**Figure H.11: Device Add Window**

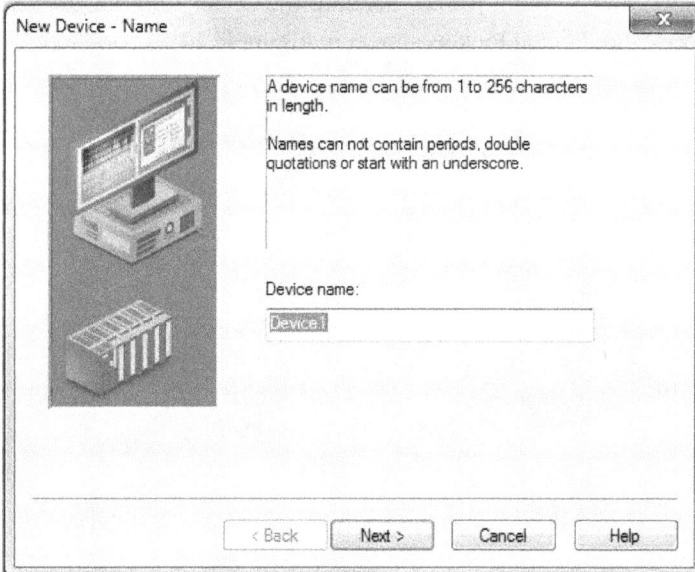

**Figure H.12: Device Name Window**

[249]

vii)    Click "Next" on the new window, and then click on the dropdown window to enter the Device ID as 1:7001 (see Figure H.13)

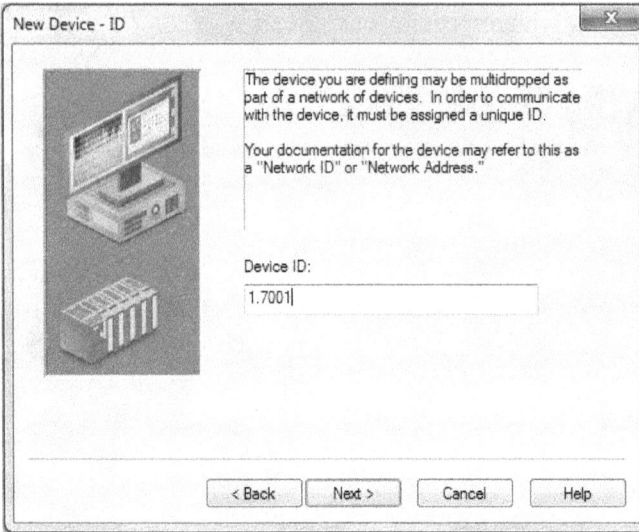

**Figure H.13: BACnet Device ID Window**

ix)    Clicking "next" until "COV" section, the select "Do not use COV". The window should then look as shown in Figure H.14.

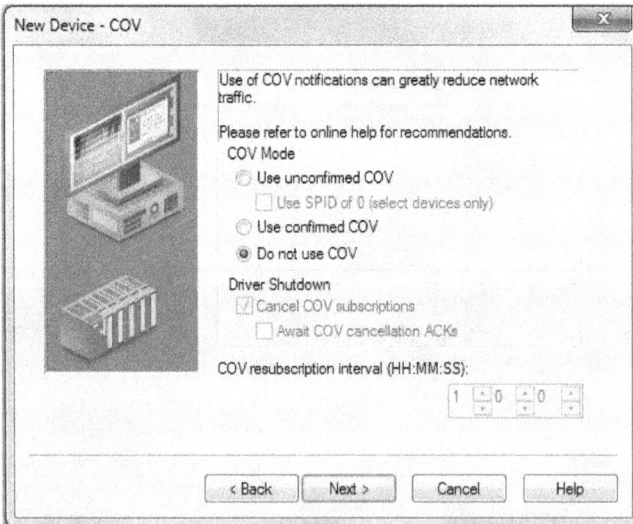

**Figure H.14: Configuration of "Change of Value" Setting**

x)   Clicking "next" until "Discovery" section, then click on "Manual Configuration" and enter the IP address as: 130.113.130.168, the window show now looks as shown in Figure H.15.

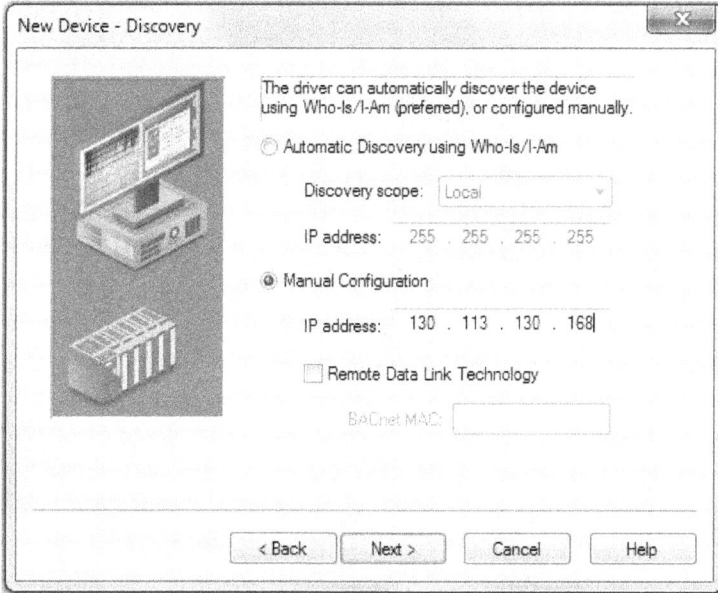

**Figure H.15: Configuration of Device Address into the BACnet OPC Channel**

xi)   Leave the remaining settings at default by clicking "next". Then click "Finish" to complete the configuration.

xii)   Save the file, then close and reopen it.

xiii)   Make sure the program is disconnect from the controller, Right click on "Device 1", click on "properties" select "Tag Import" on the window that opens (see Figure H.16), and then click on "Select Objects", and window in Figure H.17 showing the types of tags that can be imported from the devices.

**Figure H.16: Device Properties**

xii)  Check all the box identified in Figure H.17, click "Ok" " then close the window.

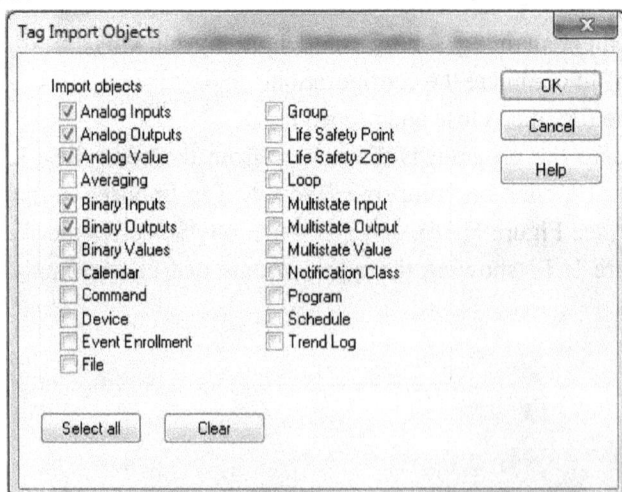

**Figure H.17: Selection of Tag Types to Import into KEPServer**

[252]

xiii) Click on "File" then "Save As" and save the file.

xiv) Start Running the program, click on "Runtime" then "Connect".

xii) Right click on "Device 1", go to "Database Creation" section, and click on "Auto Create" (see Figure H.18). The Imported tags will appear under the device as shown in Figure H.19.

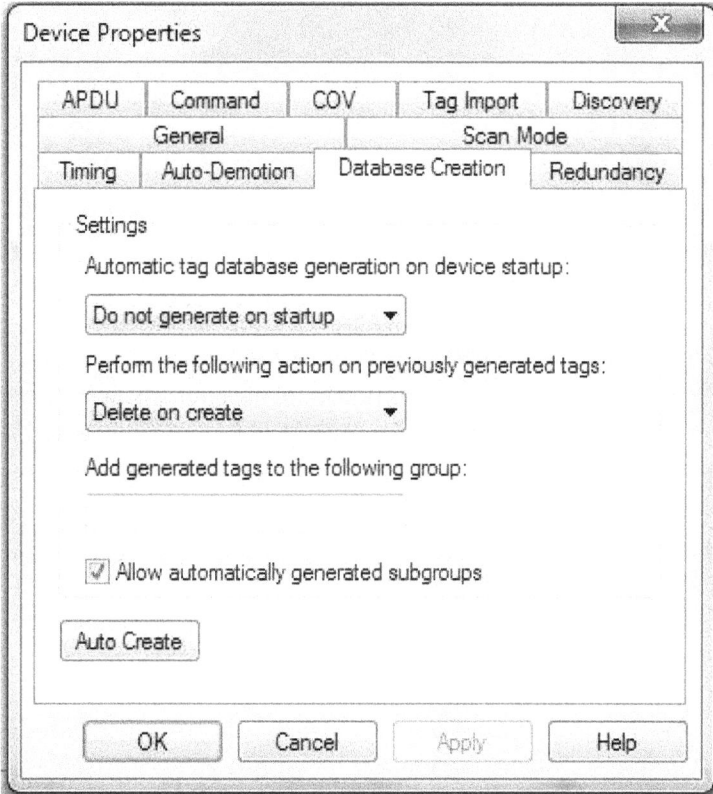

**Figure H.19: Auto Creation of Tags**

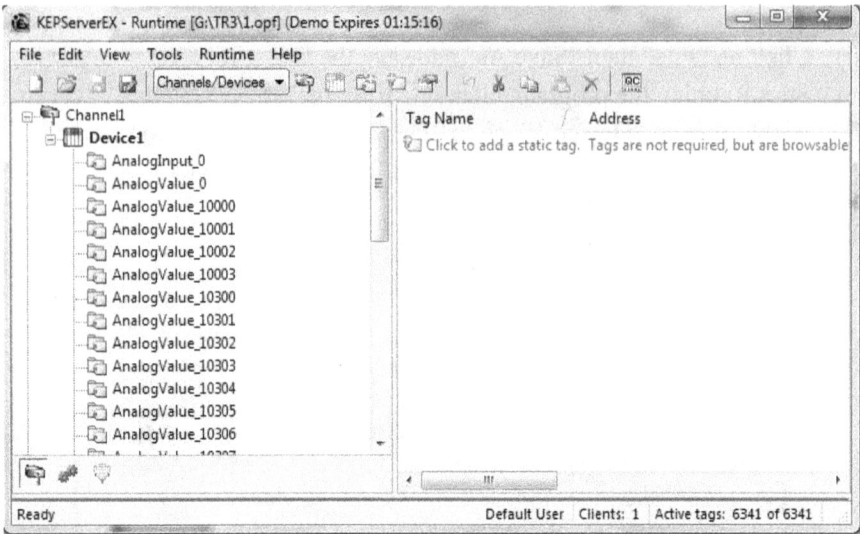

**Figure H.20: Imported Tags from PXC16 Controller**

xiii)   Click on "File" then "Save As", and save the file.

xv)     KEPServer OPC server has an inbuilt "Quick" client that enables you to view the data associated with your tags. Click on "Quick Client" to launch the quick the quick client. Then click on "Channel1.Device 1" in the main project tree window of Figure H.21. You should be able to see the controller tags include the updated values of temperature, as well as lamp and fan status.

xvi)    If KEPServer is added to OPC Datahub, PXC16 data will be accessible to the OPC client, where you can add it to the HMI, DDE or any other applications available from Datahub. Refer to lab 3B.

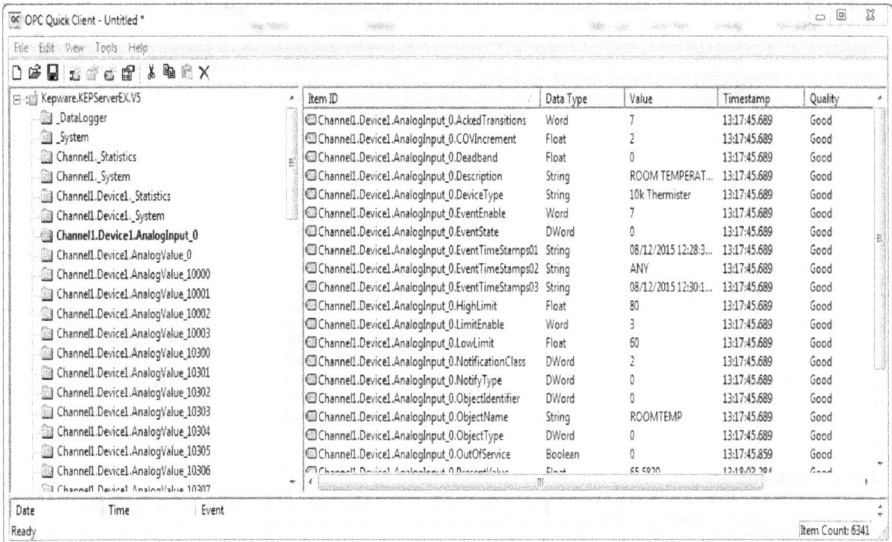

**Figure H.21: KEPServer OPC Quick Client**

# Appendix I

## Lab I: Modbus Serial Communication

### I.1    Objective

The purpose of this lab is to set up Modbus RTU communication between two PLCs, namely Productivity 3000 and Click PLC.

### I.2    Material

The following materials are required for this laboratory:

i)      Personal Computer ("PC") with Productivity Suite Programming Software
ii)     Networking Board with Productivity 3000 PLC and Click PLC

### I.3    Procedure

### I.3.1    Connect PC to LAN

i)      To view Network Connection of your computer click start →control panel→ network and sharing center (Figure I1).
ii)     Click on the network (card) connected to the lab Ethernet switch. In this case it is local area connection; then go to the network properties as follows: local area connection→properties→"Internet Protocol Version 4(TCP/IPv4)" →properties

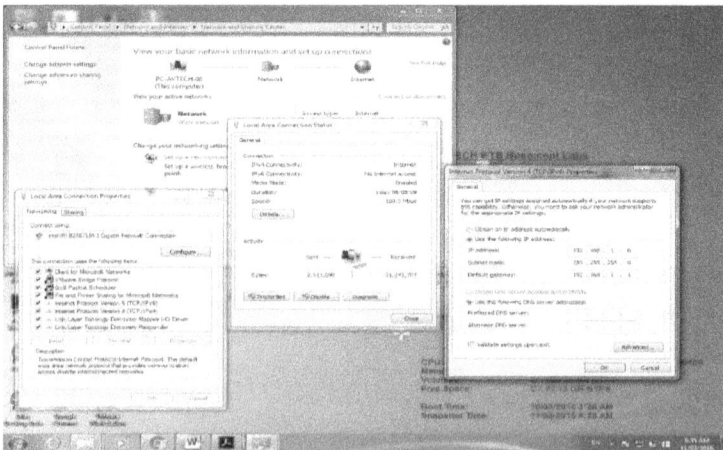

**Figure I.1: PC LAN Connection Setup**

iii) In the "Internet Protocol Version 4 (TCP/IPv4)" window, select "Use the following IP address".

iv) Enter the following settings:

- IP Address: 192.168.1. C, Subnet mask: 255.255.255.0,

- Gateway: 192.168.1.1, where PAT-C is your computer station number. For example if your computer station is PAT-6, then your IP address is 192.168.1.6.

**I.3.2 Configure PLC Ethernet Communication Parameters**

i) Plug the networking lab unit in power, connect the productivity PLC to the PC using a USB cable.

ii) Open Productivity software, click read project from PAT, choose USB

iii) 2.0, and then click connect

iv) Click "Hardware Config", window in figure 2 opens.

v) Double click on the "CPU" that is high-lightened by a rectangle in Figure I2. Window in Figure I3 opens.

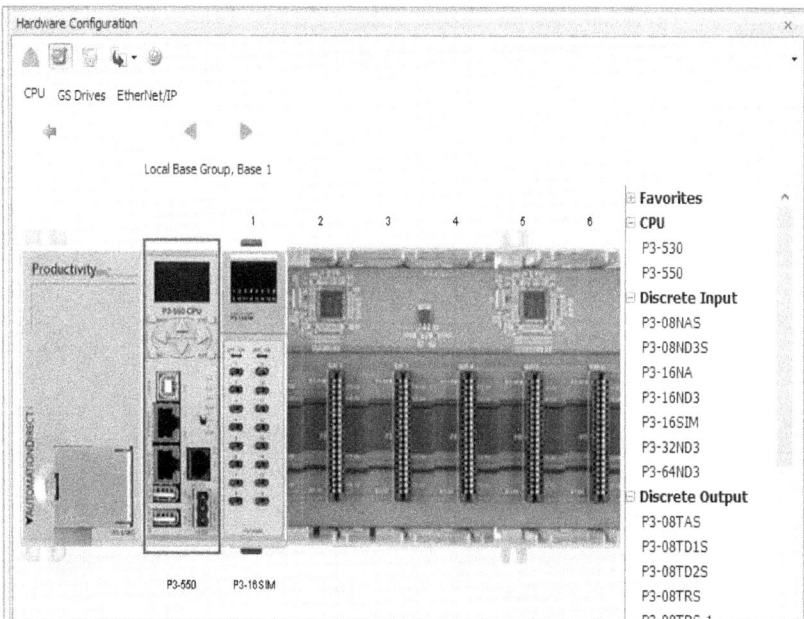

Figure I.2: CPU: Local Base Group Configuration

[257]

vi)   Click on "Ethernet Ports" in Figure I3.

vii)  For "Port Security Setting", select "Read/Write.

viii) Enter the following information:

- IP Address: 192.168.1. (X+2)C; L0X is your lab section and PAT-C is your computer station number. Do not include brackets when entering IP address.

- Subnet mask: 255.255.255.0

- Gateway: 192.168.1.1

ix)   Click "OK" on Figure I3.

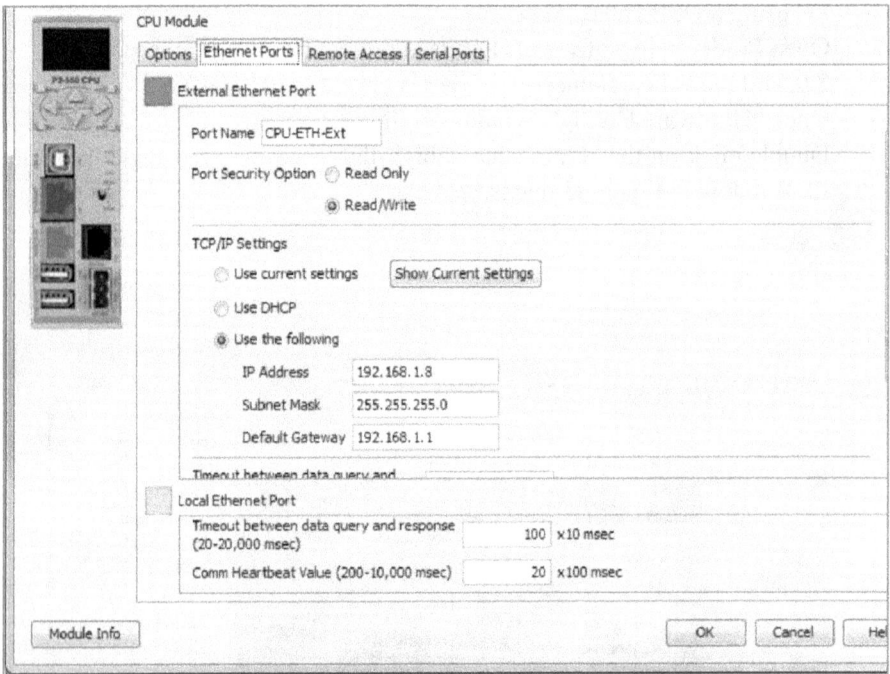

**Figure I.3: Ethernet Port Configuration**

x)    Once you have finished configuring the Ethernet communication parameters of the PLC, close productivity 3000 application, and disconnect the USB cable connecting the PLC to the PC.

xi)   Plug the Ethernet cable of the PC to the LAN switch of the networking lab unit.

[258]

xii) Open Productivity software, click read project from PAT, choose Ethernet connection, and then click connect

### I.3.3    Configure PLC Modbus Serial Communication Parameters

i)      Click on "Hardware Config" in the productivity 3000 application; window in Figure I2 opens.

ii)     Double click on the "CPU" that is high-lightened by a rectangle in Figure I2. Window in Figure I4opens.

iii)    Open "Serial Ports" tab. Under RS-232 change the Node Address to "2" to avoid both RS-485 and RS-232 from communicating on the same node (Figure I4).

iv)     Set the protocol to Modbus RTU, baud rate to 19.6K, parity to odd, data bits to 8, stop bits to 1. Note that all devices on the Modbus RTU network have been configure to these parameters.

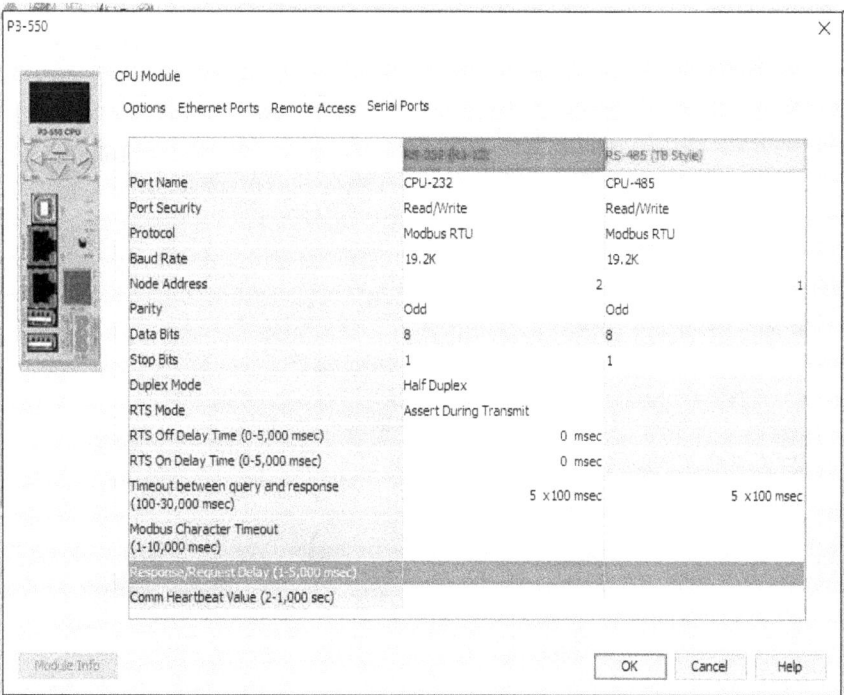

| P3-550 | | | × |
|---|---|---|---|
| CPU Module | | | |
| Options  Ethernet Ports  Remote Access  Serial Ports | | | |
| | RS-232 (RJ12 Style) | RS-485 (TB Style) | |
| Port Name | CPU-232 | CPU-485 | |
| Port Security | Read/Write | Read/Write | |
| Protocol | Modbus RTU | Modbus RTU | |
| Baud Rate | 19.2K | 19.2K | |
| Node Address | | 2 | 1 |
| Parity | Odd | Odd | |
| Data Bits | 8 | 8 | |
| Stop Bits | 1 | 1 | |
| Duplex Mode | Half Duplex | | |
| RTS Mode | Assert During Transmit | | |
| RTS Off Delay Time (0-5,000 msec) | | 0 msec | |
| RTS On Delay Time (0-5,000 msec) | | 0 msec | |
| Timeout between query and response (100-30,000 msec) | | 5 x100 msec | 5 x100 msec |
| Modbus Character Timeout (1-10,000 msec) | | | |
| Response/Request Delay (1-5,000 msec) | | | |
| Comm Heartbeat Value (2-1,000 sec) | | | |
| Module Info | | OK   Cancel   Help | |

**Figure I.4: Serial Ports Configuration**

v)      Click "OK" and close the "CPU Module" screen

## I.3.4 Modbus Write Ladder Logic Program

i) Insert a "Modbus Write" instruction at the end of the second rung (Figure I5). Reserve rung 1 for the control timer.

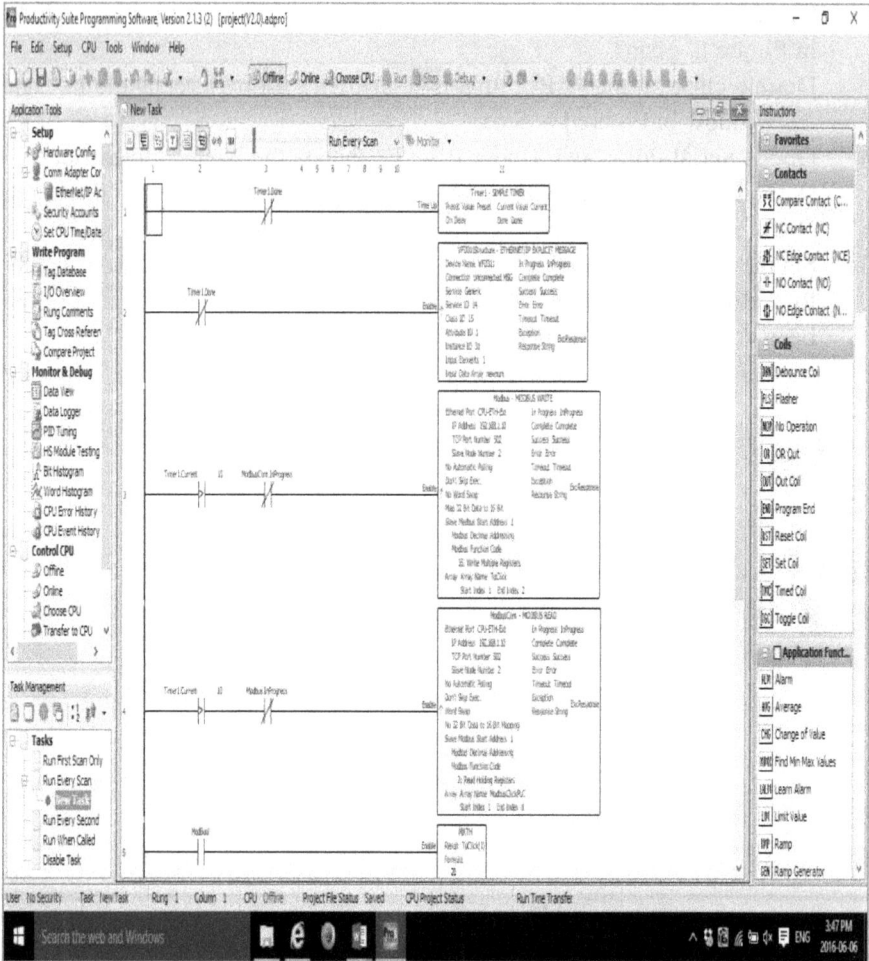

**Figure I.5: Logic for Communication Instructions**

ii) Double click on the "Modbus Write" instruction to open its configuration window in Figure I6. Select "Serial Port" and choose CPU-485, "Slave Node Number" = 2. Check "Use Structure" and name it "ModbusWrite".

[260]

iii) Enter 1 as "Slave Modbus Starting Address" and select "Modbus Decimal Addressing", and chech map 32 bit data to 16 bit.

iv) Under Tagname Mapping, Choose Write Multiple Registers as Modbus Function Code. Select "Non-Array", enter 2 as Number of Tags, and "ToClickPLC1 and ToClickPLC2" as the Tag names. Note that "ToClickPLC" are 16 bit unsigned integers. Click OK to save configuration and close Modbus Write Configuration window.

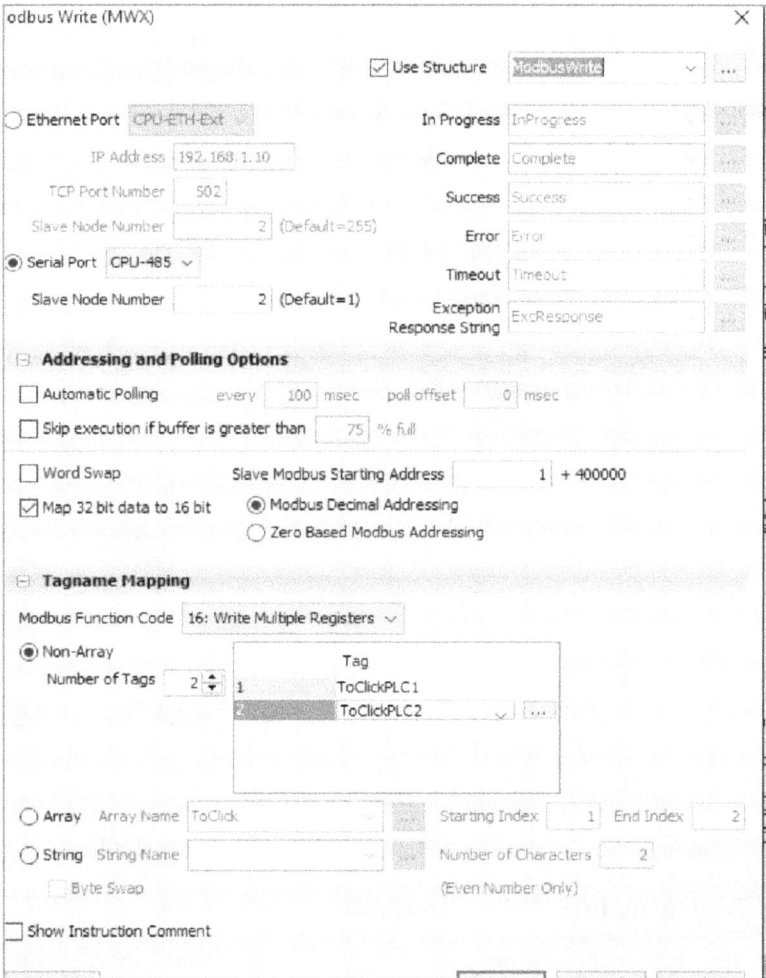

**Figure I.6: Configuration of Modbus Write Instruction**

v) Click on "monitor" in Figure 6 to add the instruction tags to "Data View", window in Figure I7 opens. Click "OK" to create a new data tab. Note its name, in this case the name is "New Task-MWX". You can change this name. If the tab has already been created, select "Append to Existing Tab", then select the tab you want to add the tags to, before clicking "Ok".

vi) Click "OK" on the Modbus instruction to close it. When prompted to create instruction tags, click "OK' to accept.

vii) Complete ladder logic program by including a timer, whose "timer.current" value is used to continually write the value of "ToClickPLC1" and "ToClickPLC2" to addresses 400001 and 400002 of the click PLC respectively. The write instruction should be enabled if timer.current<10ms.

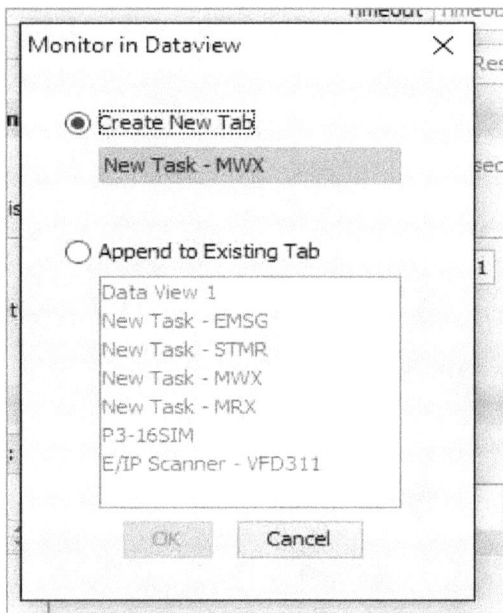

**Figure I.7: Data View Tabs**

viii) Click on "Online", then "Transfer to PAC" in the main window.

## I.3.5 Testing Modbus Write Instruction

i) After transferring the program, Click on "Data View" in the project tree window, window in Figure I8 opens

[262]

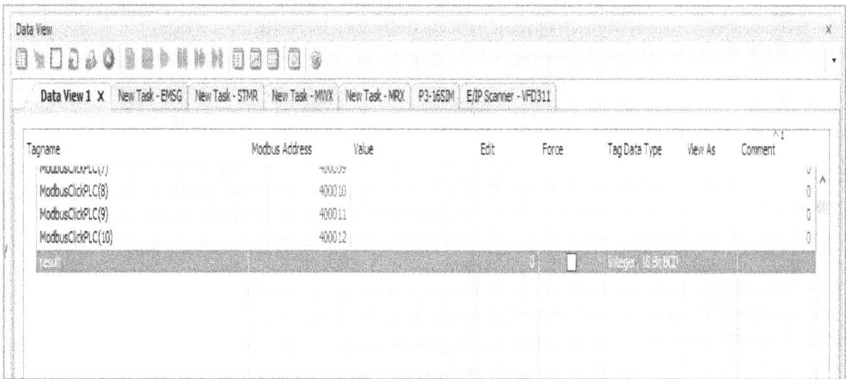

**Figure I.8: Data View**

ii) Click on "ToClickPLC1" and under "edit" type in 20 and click on "send Edit(s)".

iii) Note down your observations

iv) Enter another value into "ToClickPLC1" and note down your observations

v) Click on "ToClickPLC2" and under "edit" type in 15 and click on "send Edit(s)".

vi) Note down your observations

vii) Enter another value into "ToClickPLC2" and note down your observations

viii) Explain the observation in the lab report.

### I.3.6 Modbus Read and Ladder Logic Program

i) Insert "Modbus Read" instruction at the end of the third rung (Figure I5) of the PLC logic program.

ii) Double click on the "Modbus Read" instruction to open its configuration window. Select "Serial Port" and choose CPU-485, "Slave Node Number" = 2. Check "Use Structure" and name it "ModbusRead" (Figure I9).

iii) Enter 1 as "Slave Modbus Starting Address" and select "Modbus Decimal Addressing".

iv) Under Tag name mapping, Choose "3: Read Holding Registers" as Modbus Function Code. Select "Array", enter "FromClickPLC" as name of Tag. Enter 1 in "Starting Index" and 10 in the "End Index". This ensures that the productivity PLC reads 10 address from the click PLC, starting from 400001 to 400010 and places their values in an array FromClickPLC(i), where i=1 to 10. Note that "FromClickPLC" is an array of unsigned 16 bit integers (Figure I9).

[263]

v) Click on "monitor" in Figure I9 to add the instruction tags to "Data View", window in Figure I7 opens. Click "OK" to create a new data tab. Note its name, in this case the name is "New Task-MRX". You can change this name. If the tab has already been created, select "Append to Existing Tab", then select the tab you want to add the tags to, before clicking "Ok".

vi) Click "OK" on the Modbus instruction to close it. When prompted to create instruction tags, click "OK' to accept. You may also be prompted to increase the number of data columns for FromClickPLC tag to value equal or greater than 10. Increase the column to 15 (Figure I10).

vii) Complete ladder logic program by using the "timer.current" value to continually read the array of addresses from the click PLC. The read instruction should be enabled if timer.current>10ms.

viii) Click "OK" to save configuration and close Modbus Read Configuration window (Figure I9).

ix) Click on "Online", then "Transfer to PAC" in the main window.

x) Make sure that the PLC is in "run" mode.

xi) Click on "Monitor" in the main window of the productivity application.

xii) Go to "Data View" and enter 20 into "ToClickPLC1" to turn on Lamp1. Observe the value of each variable in the array FromClickPLC(i)xiii) Turn off Lamp 1 by entering any number other than 20 into ToClickPLC1.

xiii) You can also monitor FromClickPLC(i) by going to data "Data View" and then clicking on the + sign besides FromClickPLC tag.

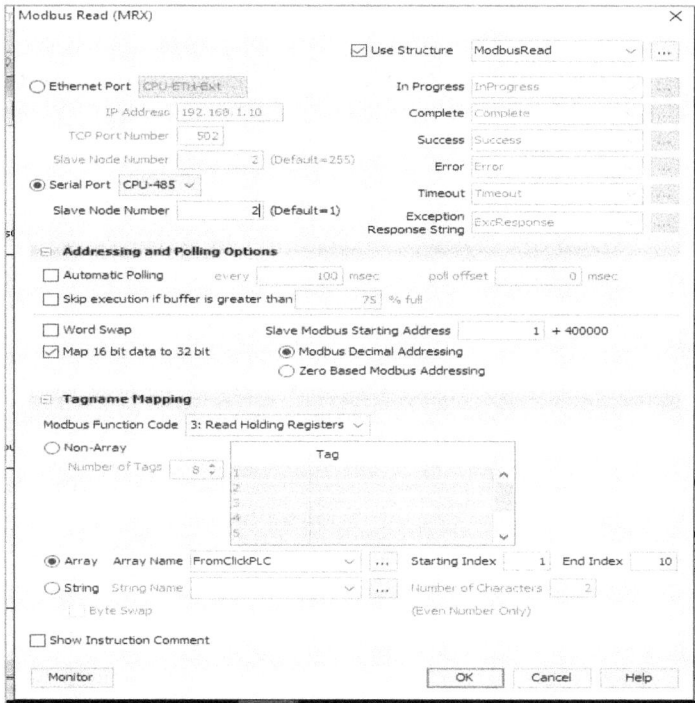

**Figure I.9: Configuration of Modbus Read Instruction**

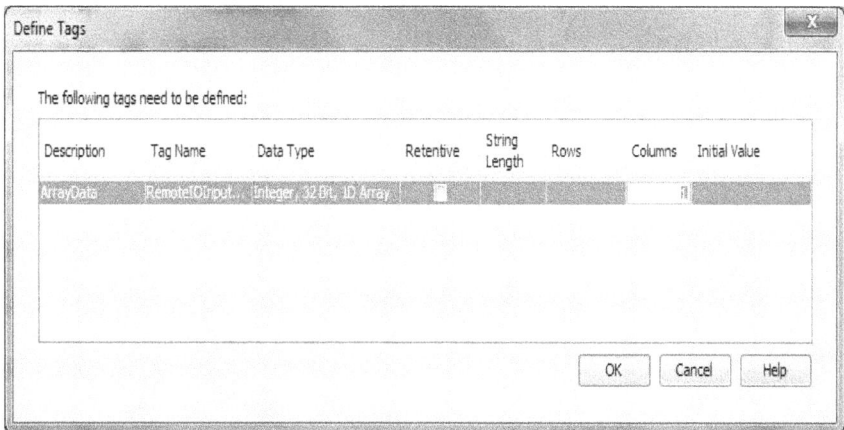

**Figure I.10: Setting Array Size**

### I.3.7    PLC Programming

i)   Tag FromClickPLC(2) is the temperature reading of a sensor installed close
     to Lamp 1 that is controlled by ToClickPLC1 tag of the Productivity 3000
     PLC. ToClickPLC2 controls Lamp 2.

ii)  Write logic that turns on Lamp 1 when the temperature is less than 28
     degrees Celsius and turns off the lamp when the temperature is above 35
     degrees.

iii) Lamp 2 should be ON when Lamp 1 is OFF, and it should be OFF when
     Lamp 1 is ON.

iv)  Use one of the inputs of the productivity 3000 PLC as a master switch to
     start/stop the Modbus communication and control of the light system. When
     the master switch is turned off, the lights should go off too. **If you are
     working online, use an internal Boolean tag instead of an input.**

# Appendix J

## Lab J: Modbus TCP Communication

### J.1    Objective

The purpose of this lab is to set up Modbus TCP communication between two PLCs, namely Productivity 3000 and Click PLC. The lab is quite similar to the Modbus RTU except that in this lab the communication is through the Ethernet network. The PLCs are connected through a Modbus TCP /RTU gateway with IP address of 192.168.1.10.

### J.2    Material

The following materials are required for this laboratory:

i)     Personal Computer ("PC") with Productivity Suite Programming Software
ii)    Networking Board with Productivity 3000 PLC and Click PLC

### J.3    Procedure

### J.3.1    Connect PC to LAN

i)     To view Network Connection of your computer click start →control panel→ network and sharing center (Figure J1).
ii)    Click on the network (card) connection connected to the lab Ethernet switch. In this case it is local area connection; then go to the network properties as follows: local area connection→properties→"Internet Protocol Version 4(TCP/IPv4)" →properties

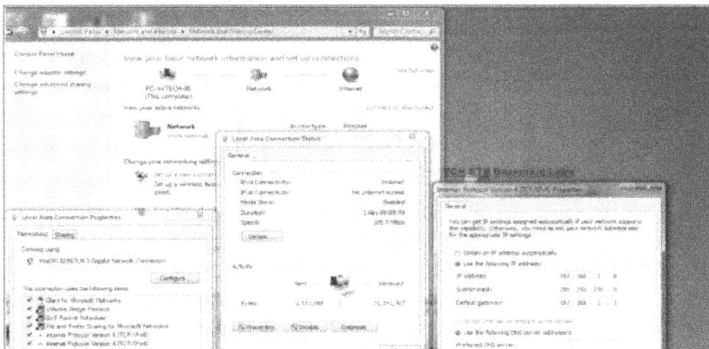

**Figure J.1: PC LAN Connection Setup**

iii)  In the "Internet Protocol Version 4 (TCP/IPv4)" window, select "Use the following IP address".

iv)  Enter the following settings:

- IP Address: 192.168.1. C, Subnet mask: 255.255.255.0,

- Gateway: 192.168.1.1, where PAT-C is your computer station number. For example if your computer station is PAT-6, then your IP address is 192.168.1.6.

### J.3.2  Configure PLC Ethernet Communication Parameters

i)  Plug the networking lab unit in power, connect the productivity PLC to the PC using a USB cable.

ii)  Open Productivity software, click read project from PAT,  choose USB 2.0, and then click connect

iii)  Click "Hardware Config", window in figure 2 opens.

iv)  Double click on the "CPU" that is high-lightened by a rectangle in Figure J2. Window in Figure J3 opens.

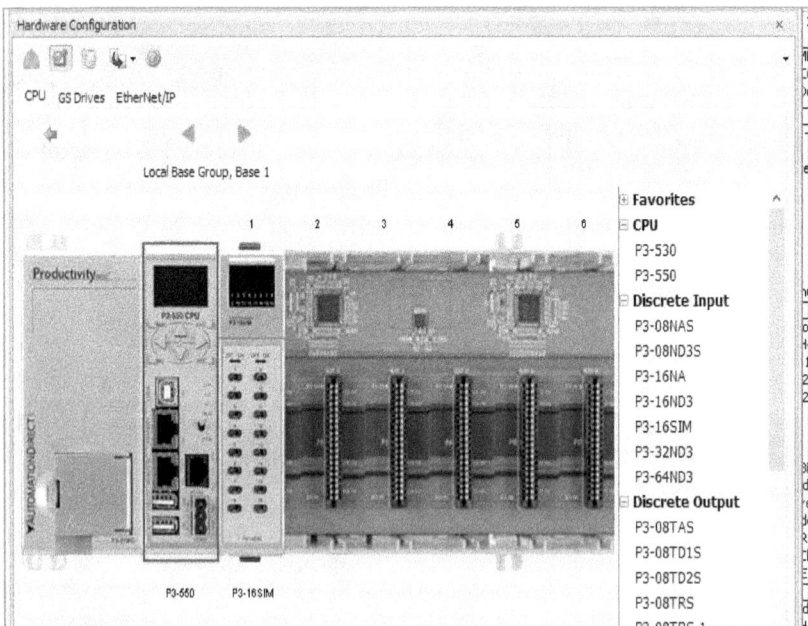

**Figure J.2: CPU: Local Base Group Configuration**

v) Click on "Ethernet Ports" in Figure J3.

vi) For "Port Security Setting", select "Read/Write.

vii) Enter the following information:

- IP Address: 192.168.1. (X+2)C; L0X is your lab section and PAT-C is your computer station number. Do not include brackets when entering IP address.

- Subnet mask: 255.255.255.0

- Gateway: 192.168.1.1

viii) Click "OK" on Figure J3.

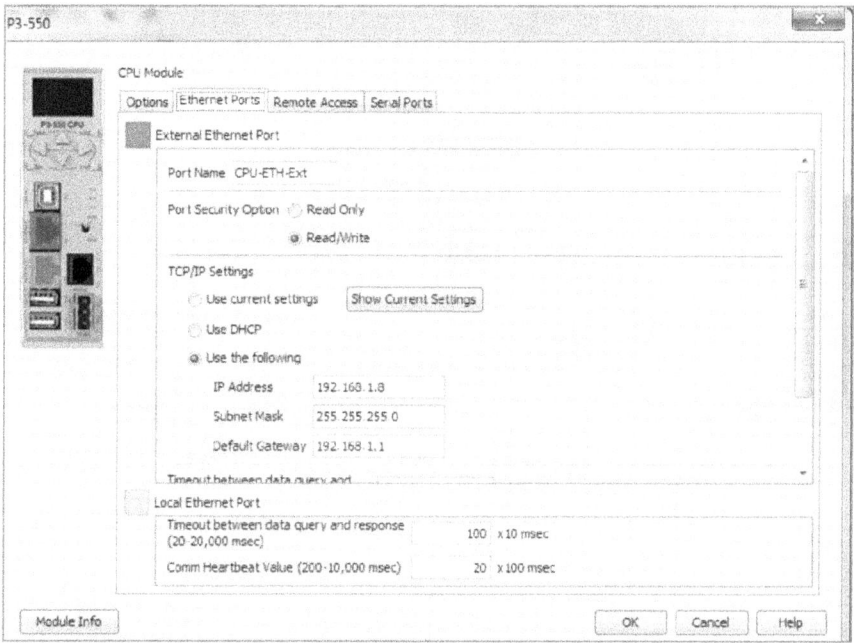

**Figure J.3: Ethernet Port Configuration**

ix) Once you have finished configuring the Ethernet communication parameters of the PLC, close productivity 3000 application, and disconnect the USB cable connecting the PLC to the PC.

x) Plug the Ethernet cable of the PC to the LAN switch of the networking lab unit.

xi)   Open Productivity software, click read project from PAT, choose Ethernet connection, and then click connect

## J.3.3   Configuration of Modbus TCP to RTU Gateway

All gateway configurations have already been made. They are provided in the lab for information only. Gateway Ethernet port is configured as shown in Figure J4, while the Modbus (serial) RTU port setting of the gateway are shown in Figure J5. They match the setting of the Click PLC. The third configuration that is made in the gateway is the device IDs for Modbus RTU Devices Connected to the Gateway (Figure J6).

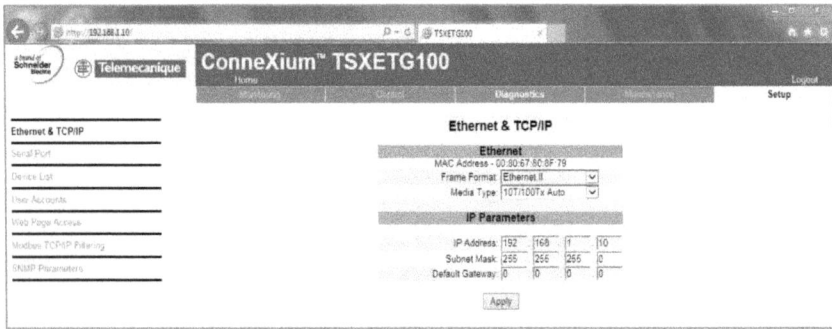

**Figure J.4: Ethernet Port Settings**

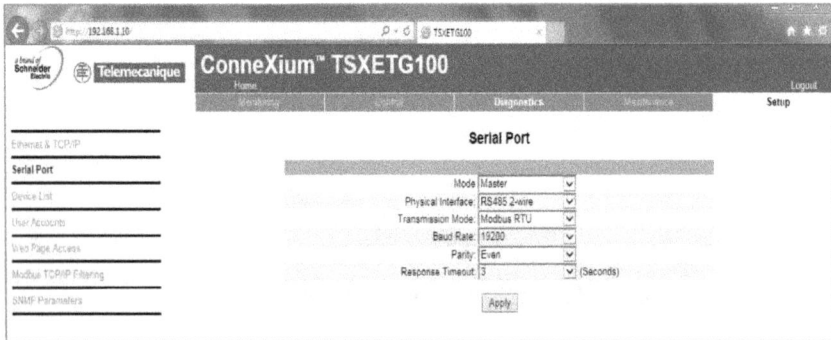

**Figure J.5: Settings of the Gateway Serial Port**

[270]

**Figure J.6: Modbus RTU Devices Connected to the Gateway**

## J.3.4    Modbus Write Ladder Logic Program

i)      Insert a "Modbus Write" instruction at the end of the second rung (Figure J7). Reserve rung 1 for the control timer.

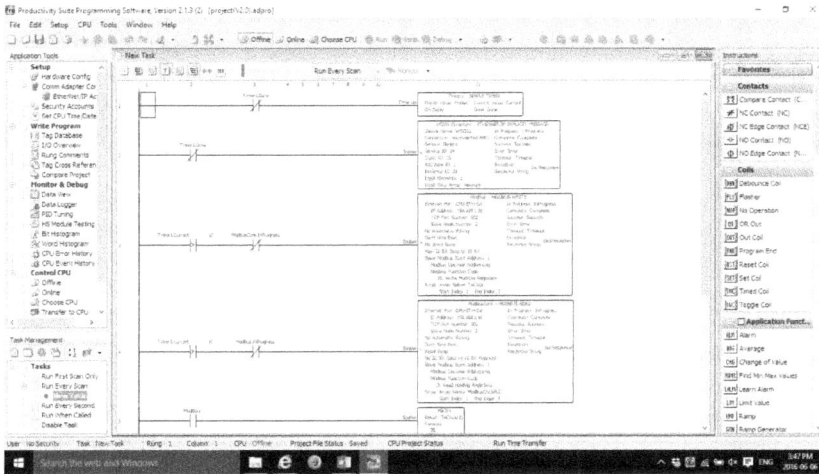

**Figure J.7: Logic for Communication Instructions**

ii)     Double click on the "Modbus Write" instruction to open its configuration window in Figure J8. Select "Ethernet Port" and choose CPU-ETHExt, IP address 192.168.1.10, "TCP Port Number" = 502, "Slave Node Number" = 2. Check "Use Structure" and name it "ModbusWrite".

iii)    Enter 1 as "Slave Modbus Starting Address" and select "Modbus Decimal Addressing", and chech map 32 bit data to 16 bit.

iv)     Under Tagname Mapping, Choose Write Multiple Registers as Modbus Function Code. Select "Non-Array", enter 2 as Number of Tags, and

[271]

v) "ToClickPLC1 and ToClickPLC2" as the Tag names. Note that "ToClickPLC" are 16 bit unsigned integers. Click OK to save configuration and close Modbus Write Configuration window.

vi) Click on "monitor" in Figure J8 to add the instruction tags to "Data View", window in Figure J9 opens. Click "OK" to create a new data tab. Note its name, in this case the name is "New Task-MWX". You can change this name. If the tab has already been created, select "Append to Existing Tab", then select the tab you want to add the tags to, before clicking "Ok".

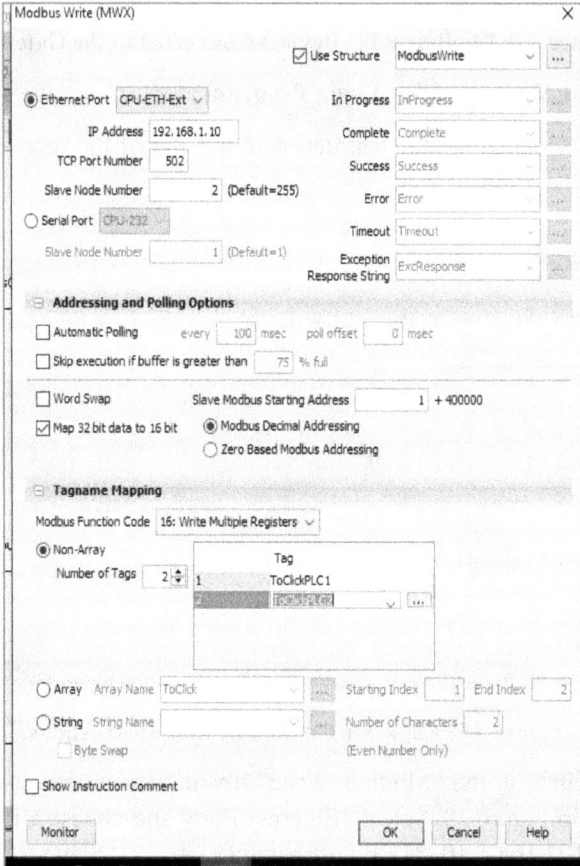

**Figure J.8: Configuration of Modbus Write Instruction**

vii) Click "OK" on the Modbus instruction to close it. When prompted to create instruction tags, click "OK' to accept.

viii) Complete ladder logic program by including a timer, whose "timer.current" value is used to continually write the value of "ToClickPLC1" and "ToClickPLC2" to addresses 400001 and 400002 of the click PLC respectively. The write instruction should be enabled if timer.current<10ms.

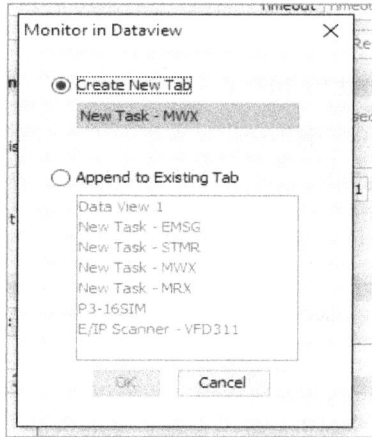

**Figure J.9: Data View Tabs**

ix) Click on "Online", then "Transfer to PAC" in the main window.

**Testing Modbus Write Instruction**

i) After transferring the program, Click on "Data View" in the project tree window, window in Figure J10 opens

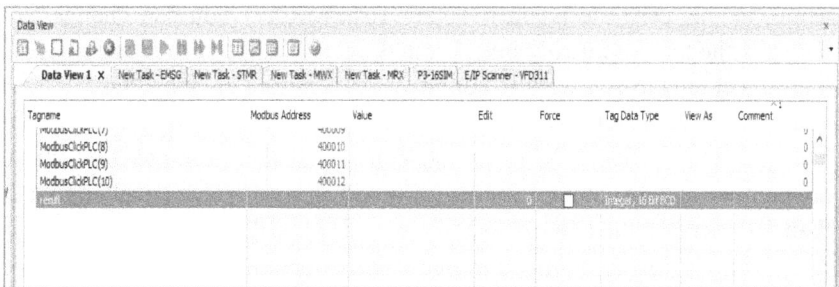

**Figure J.10: Data View**

ii) Click on "ToClickPLC1" and under "edit" type in 20 and click on "send Edit(s)".

iii) Note down your observations

iv) Enter another value into "ToClickPLC1" and note down your observations

v) Click on "ToClickPLC2" and under "edit" type in 15 and click on "send Edit(s)".

vi) Note down your observations

vii) Enter another value into "ToClickPLC2" and note down your observations

viii) Explain the observation in the lab report.

### J.3.4 Modbus Read and Ladder Logic Program

i) Insert "Modbus Read" instruction at the end of the third rung (Figure 7) of the PLC logic program.

ii) Double click on the "Modbus Read" instruction to open its configuration window in Figure 11. Select "Ethernet Port" and choose CPU-ETH-Ext, IP address 192.168.1.10, "TCP Port Number" = 502, "Slave Node Number" = 2. Check "Use Structure" and name it "ModbusRead". xvii) Enter 1 as "Slave Modbus Starting Address" and select "Modbus Decimal Addressing".

iii) Under Tag name mapping, Choose "3: Read Holding Registers" as Modbus Function Code. Select "Array", enter "FromClickPLC" as name of Tag. Enter 1 in "Starting Index" and 10 in the "End Index". This ensures that the productivity PLC reads 10 address from the click PLC, starting from 400001 to 400010 and places their values in an array FromClickPLC(i), where i=1 to 10. Note that "FromClickPLC" is an array of unsigned 16 bit integers (Figure J9).

iv) Click on "monitor" in Figure J11 to add the instruction tags to "Data View", window in Figure J9 opens. Click "OK" to create a new data tab. Note its name, in this case the name is "New Task-MRX". You can change this name. If the tab has already been created, select "Append to Existing Tab", then select the tab you want to add the tags to, before clicking "Ok".

v) Click "OK" on the Modbus instruction to close it. When prompted to create instruction tags, click "OK' to accept. You may also be prompted to increase the number of data columns for FromClickPLC tag to value equal or greater than 10. Increase the column to 15 (Figure J12). xxi) Complete ladder logic program by using the "timer.current" value to continually read the array of addresses from the click PLC. The read instruction should be enabled if timer.current>10ms.

vi) Click "OK" to save configuration and close Modbus Read Configuration window (Figure J11).

[274]

vii) Click on "Online", then "Transfer to PAC" in the main window. xxiv) Make sure that the PLC is in "run" mode. xxv) Click on "Monitor" in the main window of the productivity application.

viii) Go to "Data View" and enter 20 into "ToClickPLC1" to turn on Lamp

ix) Observe the value of each variable in the array FromClickPLC(i). xxvii) Turn off Lamp 1 by entering any number other than 20 into ToClickPLC1.

x) You can also monitor FromClickPLC(i) by going to data "Data View" and then clicking on the + sign besides FromClickPLC tag.

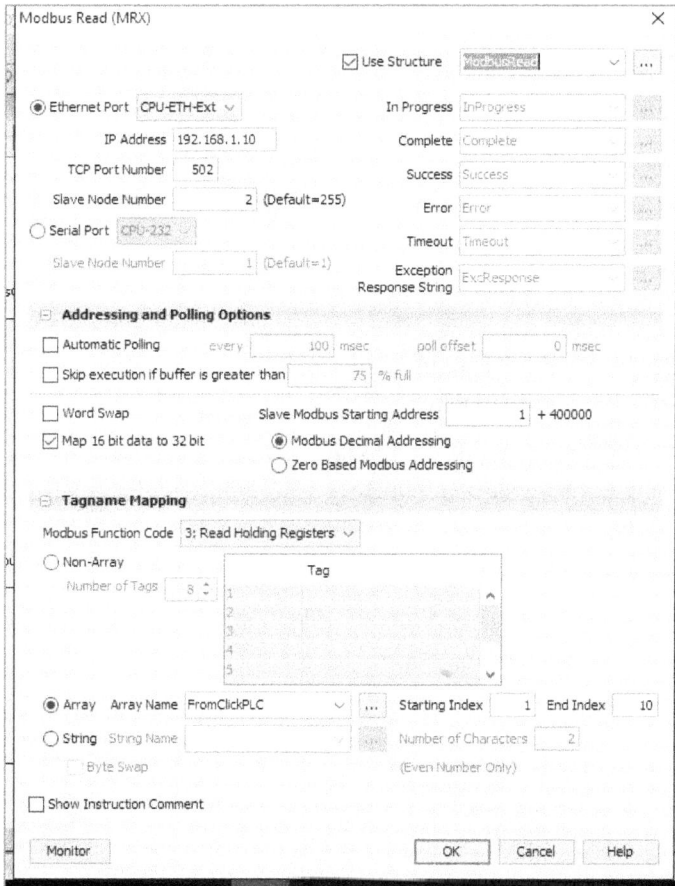

**Figure J.11: Configuration of Modbus Read Instruction**

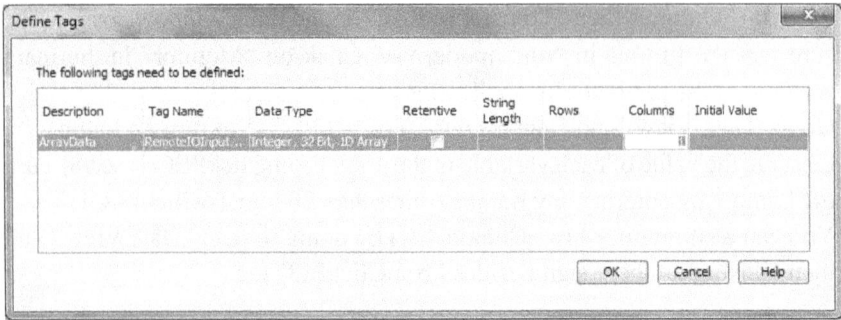

**Figure J.12: Setting Array Size**

## J.4 Exercise

Tag FromClickPLC(2) is the temperature reading of a sensor installed close to Lamp 1 that is controlled by ToClickPLC1 tag of the Productivity 3000 PLC. ToClickPLC2 controls Lamp 2.

1.  Write logic that turns on Lamp 1 when the temperature is less than 28 degrees Celsius and turns off the lamp when the temperature is above 35 degrees. vii) Lamp 2 should be ON when Lamp 1 is OFF, and it should be OFF when Lamp 1 is ON.

2.  Use one of the inputs of the productivity 3000 PLC as a master switch to start/stop the Modbus communication and control of the light system. When the master switch is turned off, the lights should go off too. If you are working online, use an internal Boolean tag instead of an input.

3.  Based on the Modbus TCP and Modbus Serial labs, compare and contrast Modbus TCP and Modbus RTU.

[276]

# Appendix K

## Lab K: Remote PLC Programming and Modbus Communication labs

### K.1    Introduction

The purpose of this set of labs is to set up Modbus communication between two PLCs, namely Productivity 3000 and Click PLC. The first lab focuses on PLC communication through Modbus TRU on RS485 network. The second lab is quite similar to the Modbus RTU one, except that in this lab the communication is through the Ethernet network. The PLCs are connected through a Modbus TCP /RTU gateway with IP address of 192.168.1.10.

These labs are based on the equipment shown in Figure 9.4 in my book titled The figure shows that the Productivity PLC is connected to the Click PLC through two ways: namely:

- Directly, using Modbus RTU RS485 network
- Indirectly, using Modbus TCP through a Modbus TCP to RTU gateway.

In the lab an Automation Direct CLICK micro PLC is configured to reading electrical parameters (voltage, current, power, and energy) from a power meter, through a Modbus RTU connection. This information is read from Click by the Productivity 3000 PLC through Modbus TRU (lab 1) or Modbus TCP (lab 2). You can program Productivity 3000 PLC to read the lamps control tags from the Click PLC and the energy data directly from the energy meter. Note that the energy meter is node 3 on the RS485 network.

### K.2    Material

The following materials are required for this laboratory:

i)    Personal Computer ("PC") with Productivity Suite Programming Software and eCatcher eWON Remote Access Software.

ii)    Networking Board with Productivity 3000 PLC and Click PLC.

iii)    eCatch Account, User Name and Password.

iv)    Camera IP Address, User Name, and Password.

## K.3   Procedure

i)    Start eCatcher software application and enter your user name and password. The lab equipment is connect to the btech account. Therefore, enter "btech'" in the account field (Figure K1).

ii)   Select any of the eWON gateways that is not in use, and click on "Connect". Once your PC is connected to the gateway minimize the eCatcher window (**do not close eCatcher**).

iii)  Open Productivity software, click read project from PAT, choose Ethernet connection, and then click connect. You are now connected to the Productivity 3000 PLC.

iv)   Go straight to Section 3.3 of the lab, because Sections 3.1 and 3.2 are already done as part of the remote access configuration.

v)    To view the lab equipment type the camera IP address in the URL of a web browser (Preferably Firefox) and enter the user name and password of the camera.

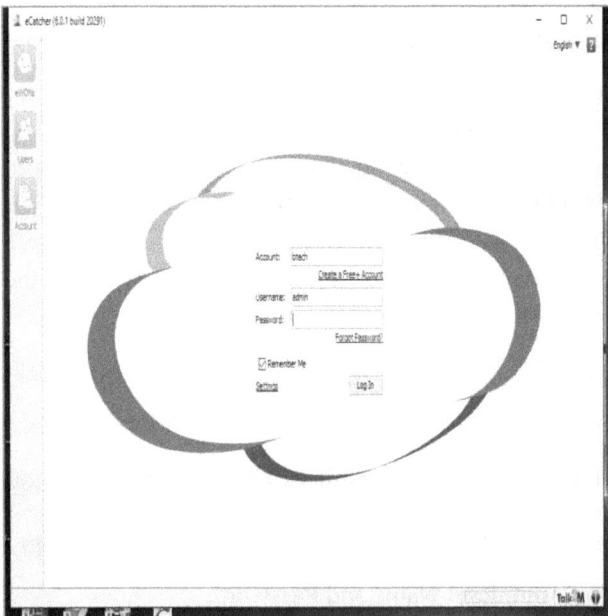

**Figure K..1: eCatcher Login Window**

[278]

# Appendix L

## Lab L: Ethernet IP Communication and Data Access Programming: PowerFle40 VFD and Productivity 3000 PLC

### L.1   Objective

The purpose of this laboratory is to program communication among Ethernet IP devices from different vendors. The laboratory helps students to learn how to program Ethernet IP implicit UDP messaging, and Ethernet IP explicit TCP connected and unconnected messaging.

### L.2   Materials

The following materials are required for this laboratory:

i)    Networks unit with PowerFlex VFD.
ii)   Productivity 3000 PLC.
iii)  Personal Computer ("the PC") with Productivity Suite Programming Software.

### L.3    Procedure

### L.3.1   Connections and Initial Setup

i)    Plug the PC RJ45 Ethernet connector to the Ethernet switch of the lab unit.
ii)   Plug the networks unit in power.
iii)  Turn on your computer

### L.3.2   Connect your Computer to the Lab LAN

i)    To view Network Connection of your computer click start →control panel→ network and sharing center →local area connection2→properties. In this lab local area connection2 is used to connect the PC to the lab unit. You may be using a different network card. Select that card in the place of local area connection2.

ii)   Once local area connection 2 properties window open, select "Internet Protocol Version 4 (TCP/IPv4)", then click properties (Figure L1).

iii)  In the "Internet Protocol Version 4 (TCP/IPv4)" window, select "Use the following IP address" option and enter the following settings: **IP Address:**

192.168.1. **C, Subnet mask**: 255.255.255.0, **Gateway**: 192.168.1.1, where PAT-**C** is your computer station number. For example if your computer station is PAT-**6**, then your IP address is 192.168.1.**6**.

### L.3.3 Assign IP Address to the VFD

i) For the PowerFlex VFD, the IP address is already set to 192.168.1.24

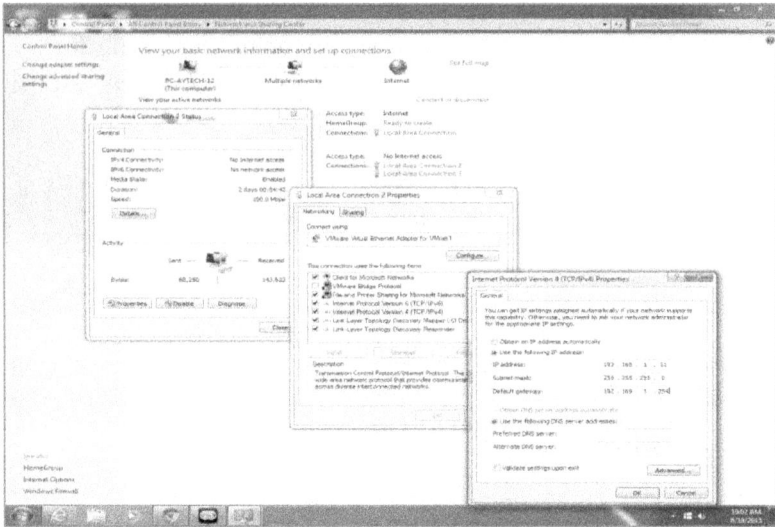

**Figure L.1: PC Network Connections Settings**

### L.3.4 Configuration of PLC Communication Parameters

i) Connect the Productivity 3000 PLC to your PC using a USB cable.

ii) Plug the power cable of the PLC in power.

iii) Open the Productivity Suit software and click on read Project from PAC Click on "Hardware Config" in the project tree window. New window opens.

iv) Click on the image under "Local Base Group" (Figure L2). New window in Figure L3 opens.

v) Click on the PLC processor highlightened by a red rectangle in Figure L3. Window in Figure L4 opens. Then clik on "Ethernet Ports".

vi) For "Port Security Setting", select "Read/Write; and for "Use the following:, enter the following communication information:

[280]

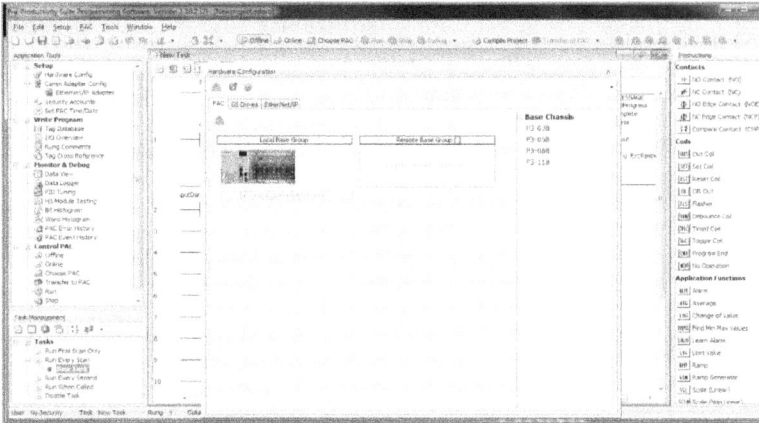

**Figure L.2: Hardware Configuration**

- IP Address: 192.168.1. (X+2)C, L0X is your lab section and PAT-C is your computer station number. Do not include brackets when entering IP address.

- Subnet mask :255.255.255.0

- Gateway: 192.168.1.1

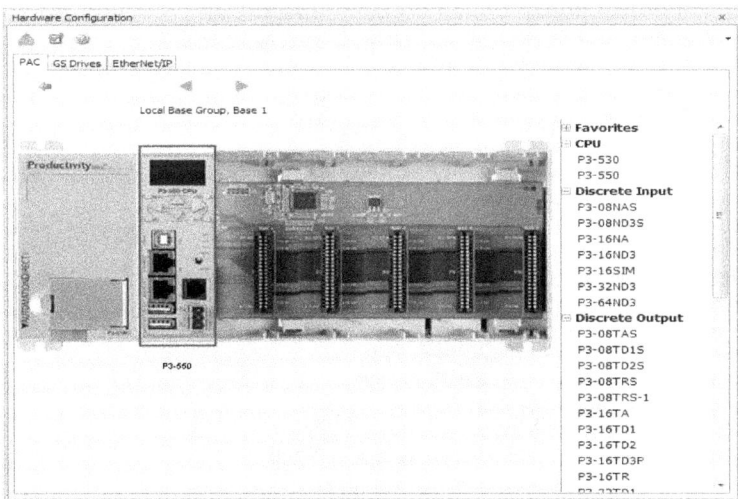

**Figure L.3: Network Configuration**

[281]

vii)   Save the screen shots on your USB memory stick to show in your report how you set the communication information.

viii)  Click "OK" and close the "Hardware Configuration" screen.

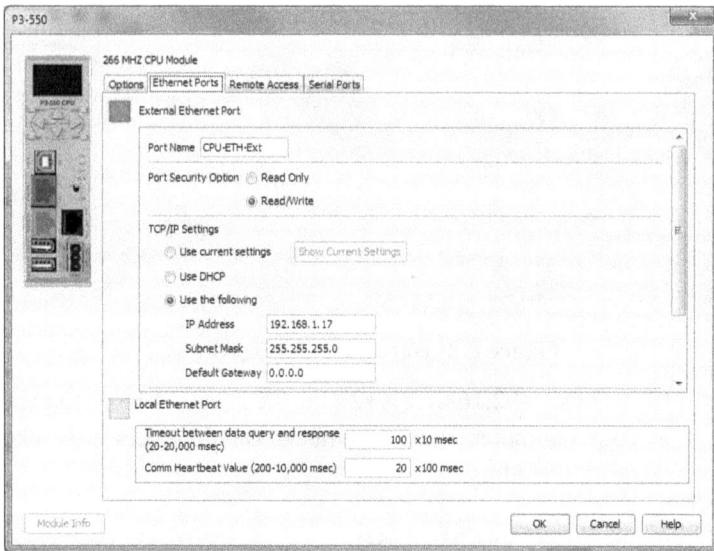

**Figure L.4: Configuration of Ethernet Port of Productivity 3000 PLC L.3.5**

## L.3.5   Configuration of Implicit IO Communication

i)   Save the screen shots on your USB memory stick throughout this lab to show in your report how you configured the communication information.

ii)   Connect the PLC to the network lab unit switch using an Ethernet cable.

iii)  In productivity 3000 software, click on "Hardware Config" in the project tree window, window in Figure L2 opens.

iv)  Click on "Ethernet/IP", new window in Figure L5 opens. In this window there are two clients already configured. Your window should be empty.

v)   Click on "Generic Client" and drag it into the window under Ethernet/IP heading. Window in Figure L6 opens.

vi)  In "Device Name", enter "VFD3XC", where X is your lab section and C is your computer station. In this lab we use "VFD311".

vii)  Enter names for "TCP Connected", Adapter Name", "Vendor ID", and "TCP Error". Every time you enter a tag you are reminded that the tag needs to be created through the window in Figure L7. Click "ok" to create the tag.

viii)  IP Address is the IP address configured into the VFD (192.168.1.24).

[282]

ix) Click on the green + sign in Figure L6 and select "Add IO Message". The window expands to look as shown in Figure L8.

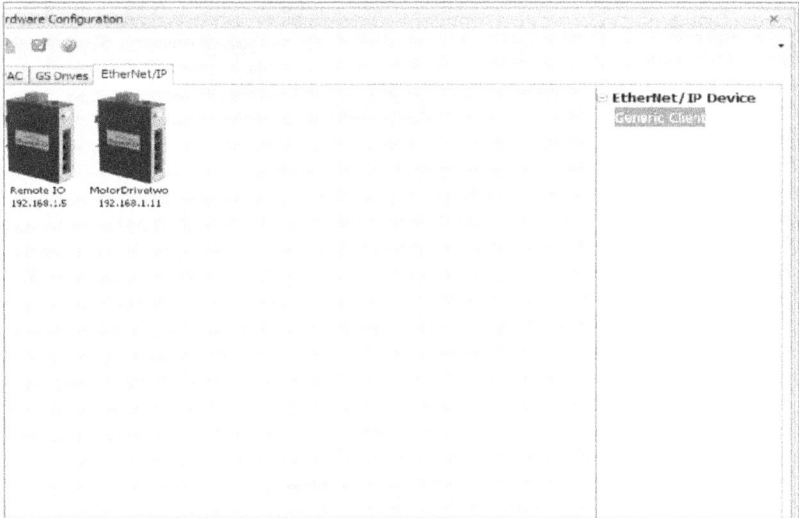

**Figure L.5: Generic Client Window**

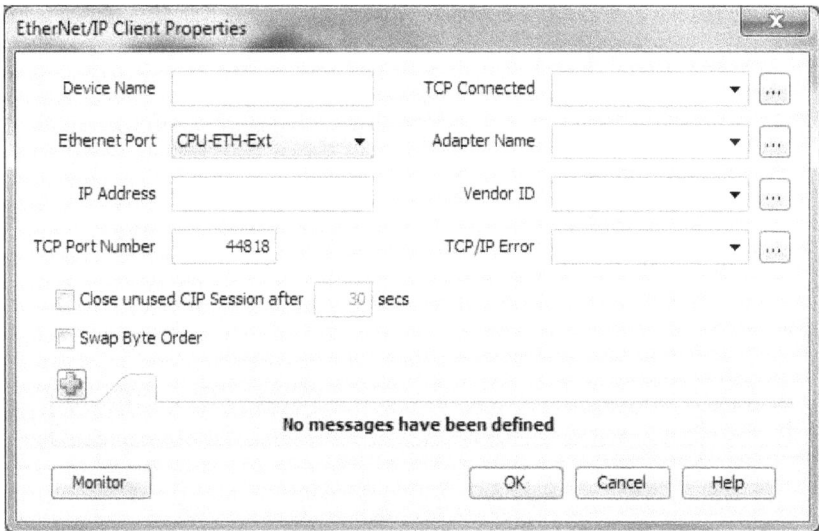

**Figure L.6: Ethernet IP Client Communication Parameters**

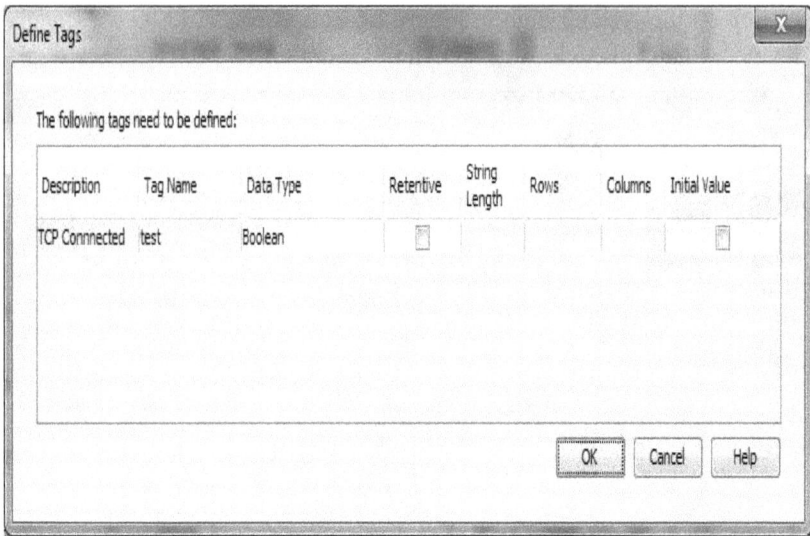

**Figure L.7: Tag Definition**

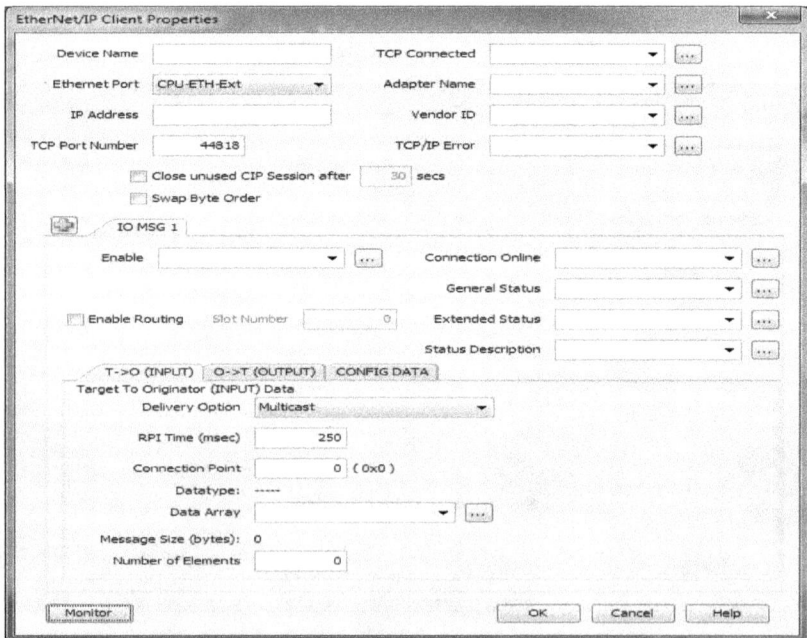

**Figure L.8: Expanded Ethernet IP Client Properties**

x)     Click on "T>O (INPUT)" and set the "Delivery Option" to "Multicast".
xi)    Leave the data update rate "RPI Time" at 250msec. Set the "Connection Point" to 1 and the" Number of Elements" to 4 (Figure L10). These parameters can be found in the VFD manual.
xii)   Your data array should be "VFDInput" with data type "Integer, 16bit Unsigned, ID Array". You will get a message that you need to increase the size of the data array. Enter a value of 5 in the window shown in Figure L9,that opens when you click "OK".

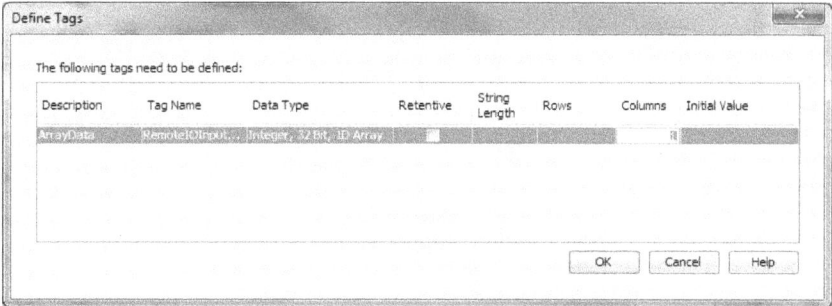

**Figure L.9: Setting Array Size**

xiii)  Click on "O>T (OUTPUT)"
xiv)   Leave the data update rate "RPI Time" at 250msec. Set the "Connection Point" to 2 and the "Number of Elements" to 2 (Figure L10). These parameters can be found in the VFD manual.
xv)    Your data array should be "VFDOutput" You will get a message that you need to increase that size of the data array. Enter a value of 4 in the window shown in Figure 9 that opens when you click "OK".
xvi)   Click on "CONFIGDATA"
xvii)  Set the "Connection Point" to 6 and the "Number of Elements" to 0 (see Figure 10). These parameters can be found in the VFD manual.
xviii) Your data array should be "VFDConfig".
xix)   Click on "Monitor" in Figure L10, so that you can be able to monitor the tags associated with this client. Window in Figure L11 opens and click "OK".

[285]

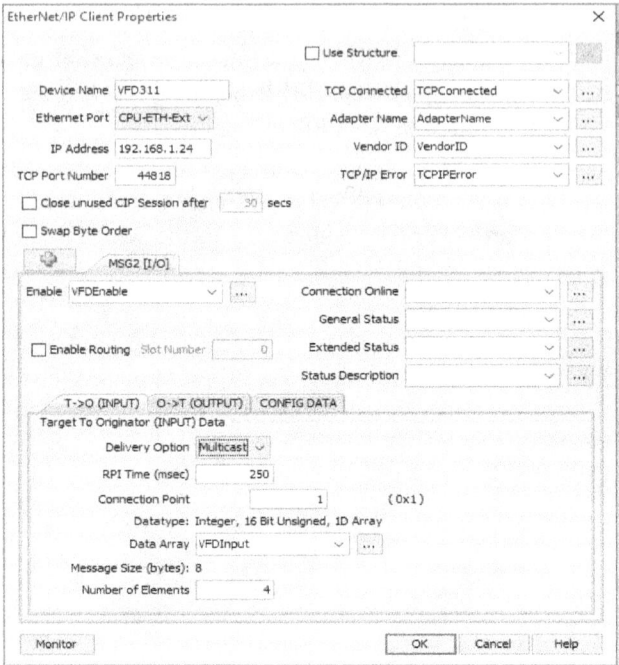

**Figure L.10: Output Parameters**

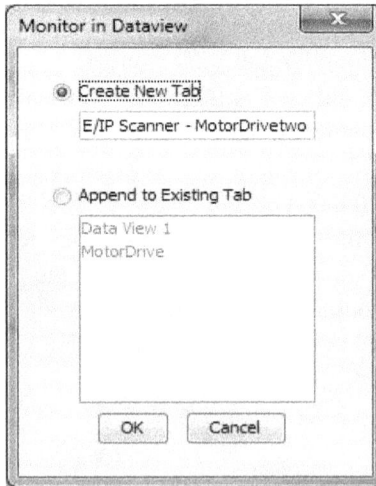

**Figure L.11: Monitor File for the Remote IO Tags**

xx)    Click "OK" in Figure L10.

[286]

xxi) Click on "Online", then "Transfer to PAC" in the main window.

xxii) After transferring the program, Click on "Data View" in the project tree window, window in Figure L12 opens. Click on the name of your device and check the box against the "Enable" tag. You can click on the + sign besides data tags to view the status of the various tags. **See step 3.5 xxix and Figure L14 on how to automatically update changes.**

xxiii) The Powerflex VFD is configured to be controlled by the PLC through the Ethernet Port. Therefore, every time it loses communications with the PLC, it goes into "Fault" state, which must be cleared before the VFD becomes operational (if you are working remotely, use the ClearFault_Command in the logic in Figure L13. In addition, the VFD communication has to be enabled every time you upload PLC configuration (see step 3.5 xxii)

xxiv) Enter the logic in Figure L13 into productivity 3000 and download to the PLC. Thereafter enable the communication between the PLC and the VFD as explained in 3.5 xxii, or by enabling the "Enable_Command" tag in the logic.

xxv) Clear the VFD fault by temporally enabling "ClearFault_Command".

xxvi) Set "Reference" to 500 and temporally enable "Start.Command" and observed the motor and the motor data (VFDInput and VFDOutput).

xxvii) Change the value of "Reference" and observe the motor and its associated data.

xxviii) Temporally enable "Stop_Command".

xxix) In your report explain the communication configuration, the logic, and the actions of the motor.

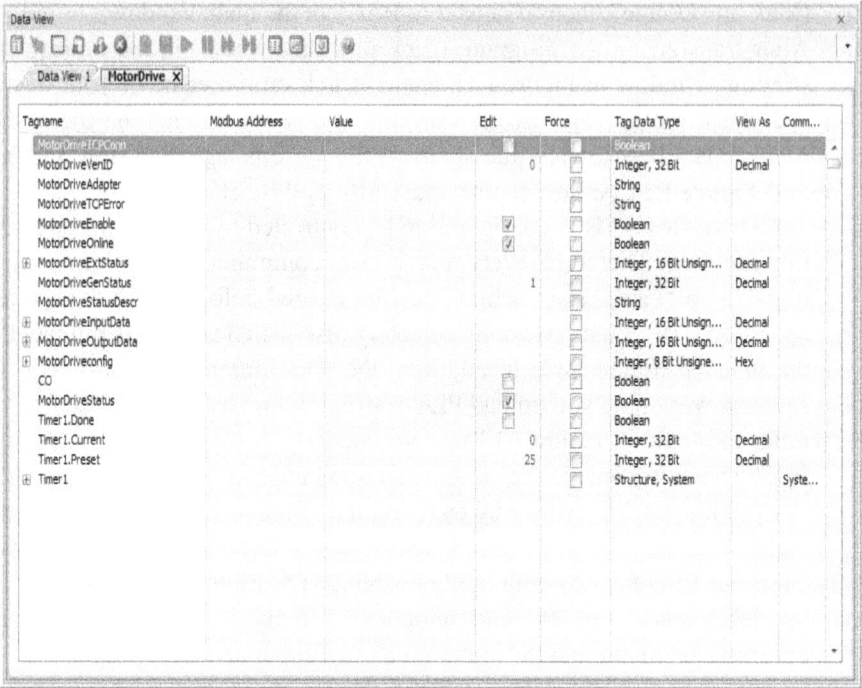

**Figure L.12: Data View**

xxx) To enable data change to be updated automatically click on "data options", widow below opens, then check" Enable Auto Edits", then click "ok" (Figure L14).

[288]

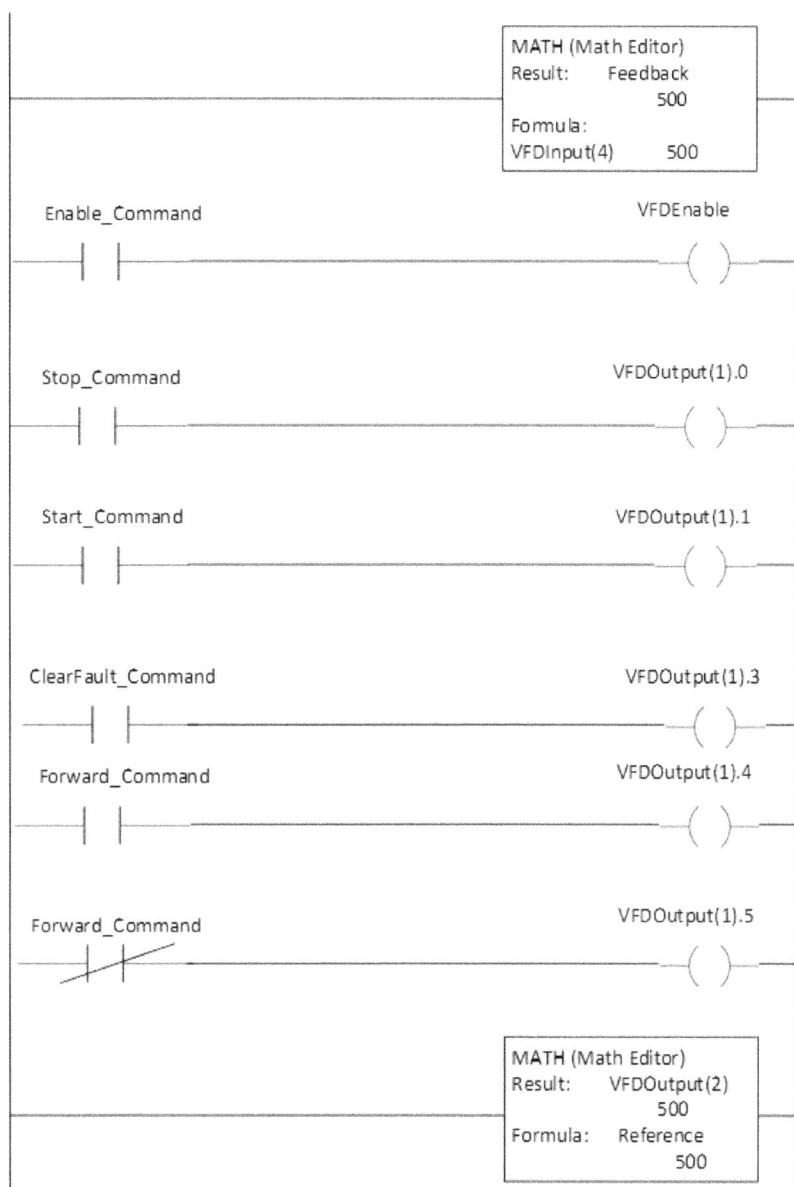

**Figure L.13: Motor Control Logic**

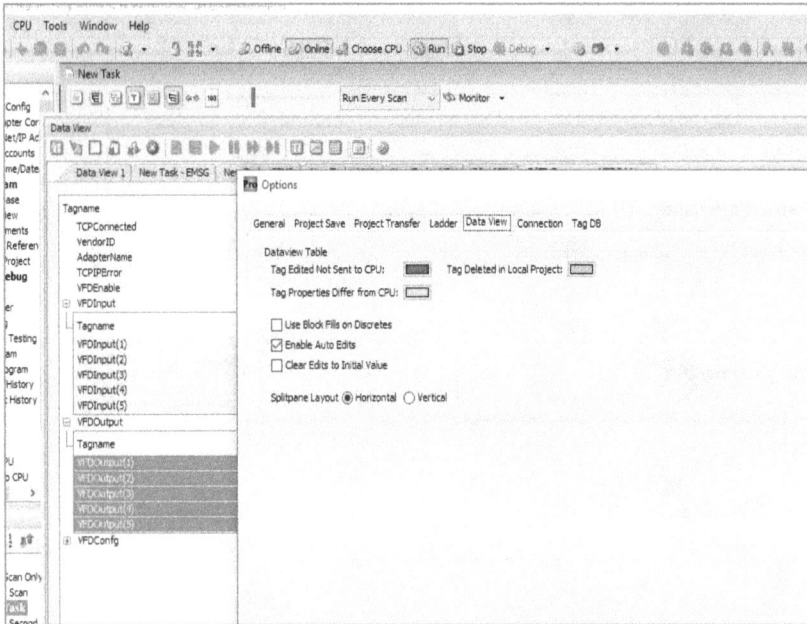

**Figure L.14: Updating Changes Automatically**

### L.3.6 Configuration of Explicit Unconnected Ethernet IP Communication

i) The VFD client has already been configure through steps 3.5 iii) to 3.5 viii.

ii) Click on "End" on the last rung in the programming window, and then click on the "Ethernet/IP Explicit Message" instruction. Window in Figure L15 opens. For "Device Name", select "VFDCX"; in our case we used VFD311. "Connection", leave unchanged; "Service", **SELECT** "Generic".

iii) Enter the following parameters, "Service ID = 14", Class ID = 15", "Attribute ID = 1", "Instance ID = 3. These parameters can be found in the VFD manual. Attribute 1 is the IP address and instance 3 of the attribute is the first octet of the address.

iv) Check "Enable Input", enter the name of your data point, in this case *IPAddressOctet*, and set the "Number Elements" to 1.

v) Click "Monitor", then "Ok" to create monitored data, then click "Ok" to close the Ethernet IP Explicit Message configuration.

vi) The Message instruction is executed only when the rung status changes from false to true. If you want to continuously update data, you need to use a timer instruction as shown in Figure L16.

[290]

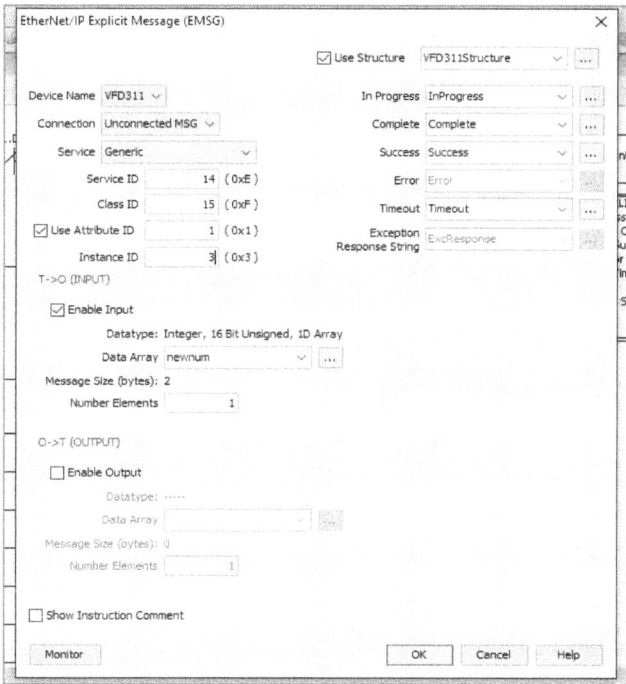

**Figure L.15: Ethernet IP Explicit Message Configuration**

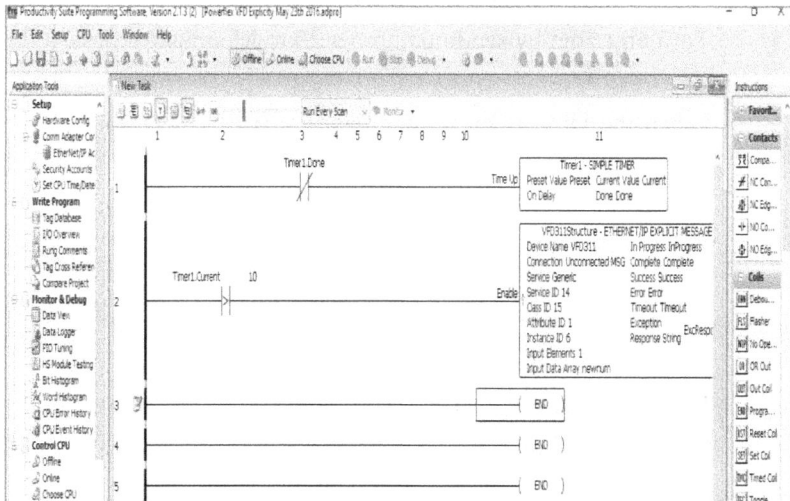

**Figure L.16: Explicit Messaging Logic**

vii)  Note that the *time.reset* value is entered by going to "Data View" and creating the monitoring data set.

viii)  Download the PLC configuration, and put the PLC in "run" mode.

ix)  What is the value of *IPAddressOctet*?

x)  Change the instance ID to 4, download the configuration to the PLC, and put it in "run" mode.

xi)  Note down the value of *IPAddressOctet*.

xii)  Repeat steps xi and xii with instance ID of 5 and then 6.

xiii)  In your report explain a situation where the communication in this section of the lab would be desirable.

### L.3.7  PLC Programming Exercise

i)  Save the screen shots on your USB memory stick throughout this lab section to show in your report how your logic.

ii)  Write the state machine, flow chart, and logic that accomplishes the following:

- The initial state of the system is Lamp is off, temperature is less than 28 deg., and motor is off.

- Read temperature continuously from the click PLC using Modbus TCP through a gateway with IP address 192.168.1.10, "TCP Port Number" = 502, and "Slave Node Number" = 2.

- Turn on Lamp by sending number 20 to Modbus Address 400001 of the click PLC.

- When the lamp is on the temperature rises. When the temperature reaches 30 degrees centigrade, turn on the motor.

- The motor speed should be proportional to the temperature, such that when the temperature is 28, the speed should be 0; and temperature of 38 degrees, should correspond to the speed of 60 Hz.

- When the temperature reaches 38 deg. The lamp should go off. But the motor should stay on for 30 seconds.

- Lamp should not go on again until the system returns to initial state.

- The system should return to initial state once the motor has stopped.

iii)  Enter the logic into the PLC IDE and download to the PLC.

iv)  Demonstrate your results to the instructor.

# Appendix M

## Lab M: Ethernet IP Communication and Data Access Programming: Powerfle40 VFD and Compactlogix or Contrologix PLC

### M.1 Objective

The purpose of this laboratory is to program communication between Rockwell Automation Ethernet IP devices. The laboratory helps students to learn how to program Ethernet IP implicit UDP messaging, and Ethernet IP explicit TCP unconnected messaging.

### M.2 Materials

The following materials are required for this laboratory:

i) Networks unit with Powerflex VFD.
ii) Compactlogix or Contrologix PLC
iii) Personal Computer ("the PC") with RSLogix5000 Software.

### M.3 Procedure

#### M.3.1 Connections and Initial Setup

i) Plug the PC RJ45 Ethernet connector to the Ethernet switch of the lab unit.
ii) Plug the networks unit in power.
iii) Turn on your computer

#### M.3.2 Connect your Computer to the Lab LAN

i) To view Network Connection of your computer click start →control panel→ network and sharing center →local area connection2→properties. In this lab "local area connection 2" is used to connect the PC to the lab unit. You may be using a different network card. Select that card in the place of "local area connection 2".
ii) Once "local area connection 2" properties window open, select "Internet Protocol Version 4 (TCP/IPv4)", then click properties (Figure M1).
iii) In the "Internet Protocol Version 4 (TCP/IPv4)" window, select "Use the following IP address" option and enter the following settings: **IP Address:** 192.168.1. **C, Subnet mask:** 255.255.255.0, **Gateway:** 192.168.1.1, where PAT-**C** is your computer station number. For example if your computer

station is PAT-**6**, then your IP address is 192.168.1.**6**. Note that any other IP address within the range can be used.

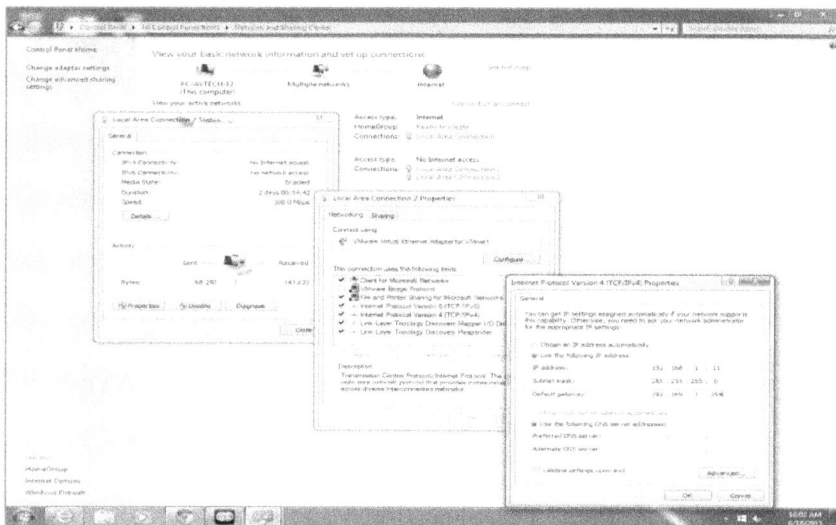

**Figure M.1: PC Network Connections Settings**

## M.3.3 Assign IP Address to the VFD

i)     For the Powerflex VFD, the IP address is already set to 192.168.1.24

## M.3.4 Configuration of PLC Communication Parameters

i)     In this lab it is assumed that the PLC has already been configured for communication, and it is ready to receive logic from the IDE (RSLogix5000).

## M.3.5 Configuration of Implicit IO Communication

i)     Save the screen shots on your USB memory stick throughout this lab to show in your report how you configured the communication information.

ii)    Connect the PLC to the network lab unit switch using an Ethernet cable.

iii)   In the project tree of RSLogix5000, right-click on the Ethernet connection (i.e. port, scanner, or bridge). In our example, we right-click on Ethernet (Figure M2), then click on "New Module". Window in Figure M3 opens.

iv)    In Figure M3 click on "Ethernet Module (Generic Ethernet Module)" followed by clicking on "Create". Module properties dialog box in Figure

[295]

M4 opens. Fill the box as follows: Name, type in VFD311 (or any name to identify the drive), Comm. Format, select **Data – INT**, this setting formats the data to 16-bit words; and IP Address is the IP address of the drive. Enter the connection parameters as shown in Figure M4. These parameters are available in the VFD manual.

v)    Click "OK", and window in Figure M5 opens.

vi)   Check "Use Unicast over EtherNet IP" (Figure M5), then click "OK" to complete the configuration.

vii)  Download the configuration to the PLC as follows: Select **Communications > Download**. Then click on "download" in the **download** dialog box.

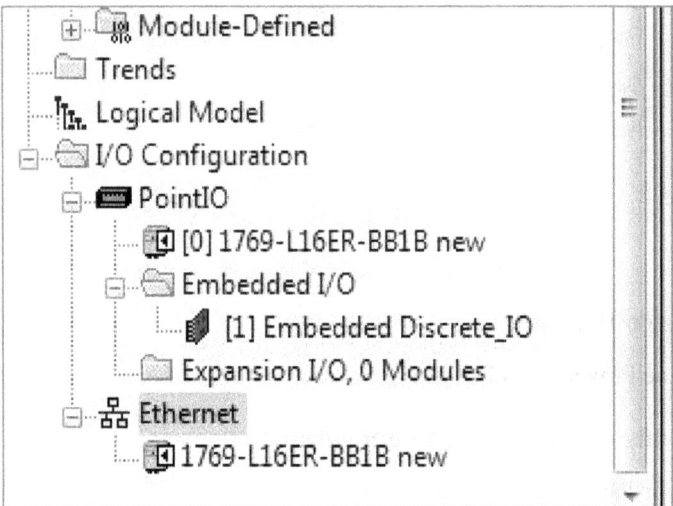

**Figure M.2: Right Click on Ethernet**

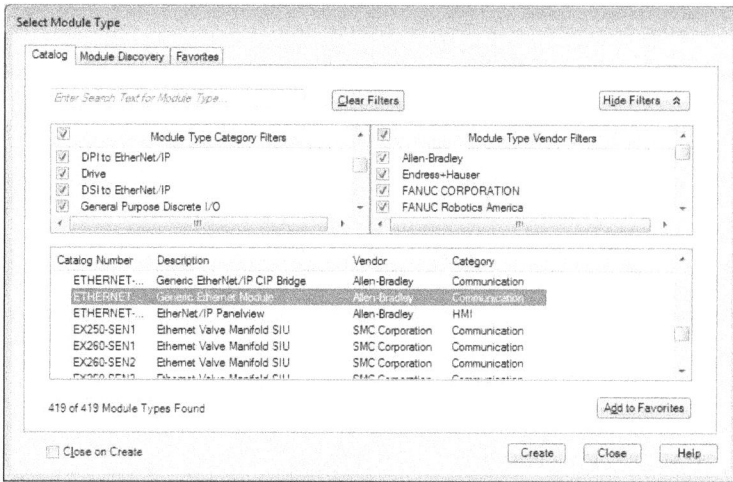

**Figure M.3: Module Type**

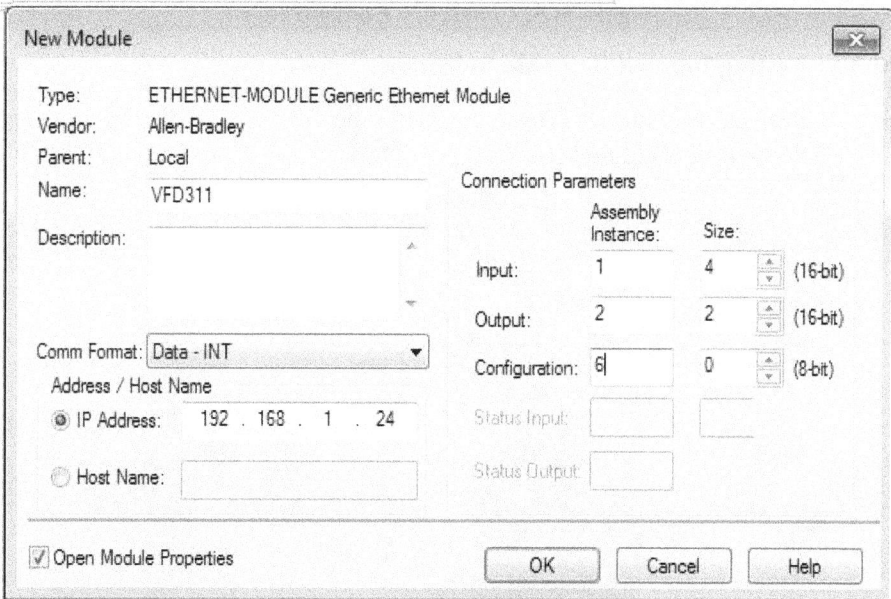

**Figure M.4: Module properties dialog box**

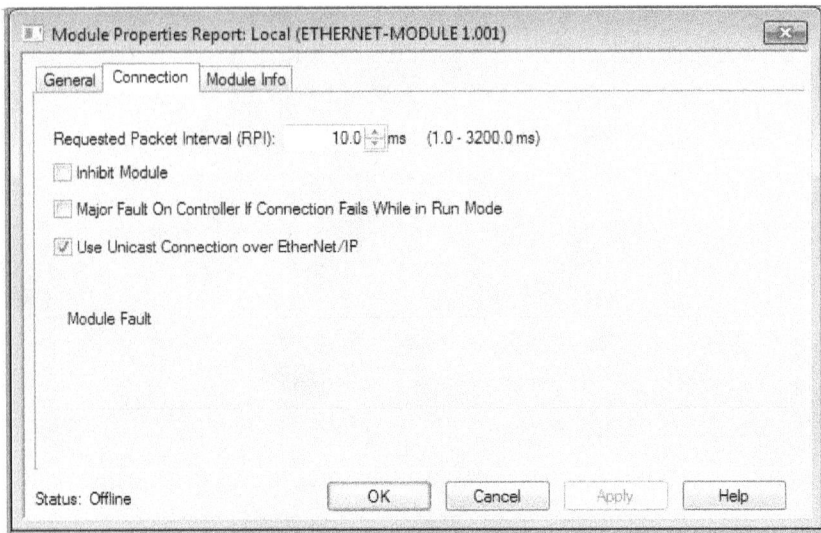

**Figure M.5: Connection Configuration**

## M.3.6 Testing Implicit IO Communication

i)  Click on "Controller Tags" to view the VFD tags that are created automatically (Figure M6).

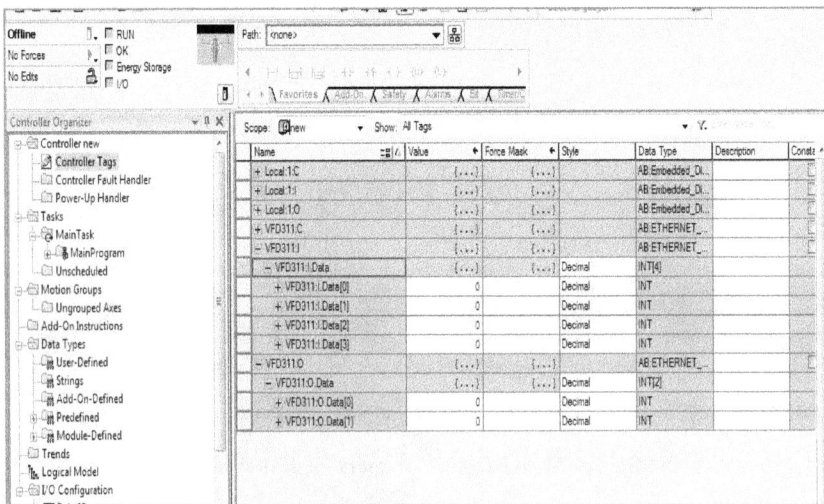

**Figure M.6: VFD Tags**

ii) Open the "Tasks" item in the project tree by clicking on "+". Then double click on "MainRoutine" to open the programming window.

iii) The Powerflex VFD is configured to be controlled by the PLC through the Ethernet Port. Therefore, every time it loses communications with the PLC, it goes into "Fault" state, which must be cleared before the VFD becomes operational (if you are working remotely, use the ClearFault_Command in the logic in Figure M7.

iv) Enter the logic in Figure M7 into RSLogix5000 IDE and download to the PLC.

v) Clear the VFD fault by temporally enabling "ClearFault_Command".

vi) Set "Reference" to 500 and temporally enable "Start_Command". Observed the motor and the motor data (VFDInput and VFDOutput).

vii) Change the value of "Reference" and observe the motor and its associated data.

viii) Temporally enable "Stop_Command". ix) Test the "Forward" and "Reverse" commands.

ix) In your report explain the communication configuration, the logic, and the actions of the motor.

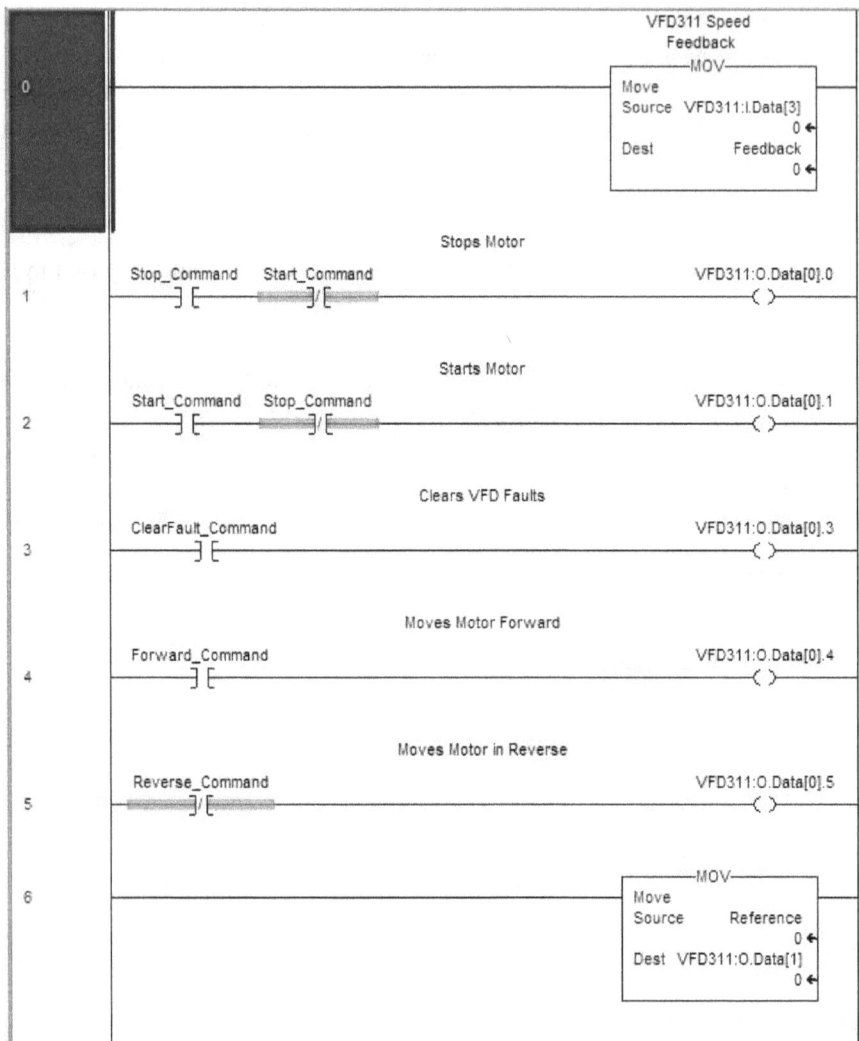

**Figure M.7: Motor Control Logic**

## M.3.7 Configuration of Explicit Unconnected Ethernet IP Communication

i)    Insert a message instruction on a rung as shown in Figure M8 click on the "?" in front of "Message Control" and type in "VFD311MSG".

ii)   Right click on "VFD311MSG" and create the message type tag at the controller level.

[300]

iii) Double click on the icon in front of "VFD311MSG" and the window in Figure M9 opens.

iv) Fill out the window in Figure M9as follows:

1- **Message Type:** Should be set to CIP Generic.

2 - **Service Type:** The service type indicates the service (for example, Get Attribute Single or Set Attribute Single) that you want to perform. Available services depend on the class and instance that you are using. In this case select "Get Attribute Single".

3 - **Service Code:** Code e, associated with "Get Attribute Single" is automatically entered.

4 – **Class:** The class is an Ethernet IP class. In this case enter f, hexadecimal for 15. This is the device parameters class.

5 – **Instance:** This is an instance (or object) of an Ethernet IP class. In this case enter 3, which is the instance "first octet" of object "IP address".

6- **Attribute:** The attribute is a class or instance attribute. In this case enter 1. Note that Attribute 1 is the IP address of the VFD,

7 - **Source Element:** This box contains the name of the tag for any service data to be sent from the PLC to the VFD. In this case it is blanked out because we are reading from the VFD.

8 - **Source Length:** This box contains the number of bytes of service data to be sent in the message. In this case it is blanked out.

9 – **Destination:** This box contains the name of the tag that will receive service response data from the drive. Click on "New Tag" and type "IPAddressOctet"in the name box of the window that opens. Set data type to INT (Integer). The click on the arrow facing would and select"IPAddressOctet" from the tags.

10 – **Path:** The path is the route that the message will follow. Click "**Browse**" to find the path or type in the name of an adapter that you previously mapped.

11- **Name:** The name for the message. In this case "VFD311MSG" was entered during message configuration.

The Message instruction is executed only when the rung status changes from false to true. If you want to continuously update data, you need to use a timer to control the VFD_Read tag.

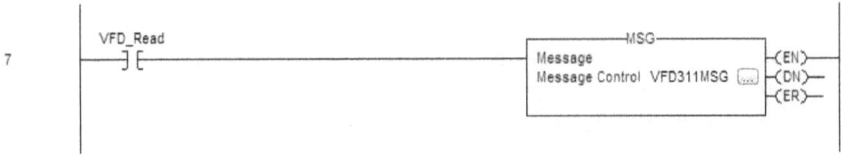

**Figure M.8: Ethernet IP Explicit Message Instruction**

v)     Download the PLC configuration, and put the PLC in "run" mode.

vi)     What is the value of IPAddressOctet?

vii)     Change the instance ID to 4, download the configuration to the PLC, and put it in "run" mode.

viii)     Note down the value of IPAddressOctet.

ix)     Repeat steps xi and xii with instance ID of 5 and then 6.

x)     In your report explain a situation where the communication in this section of the lab would be desirable.

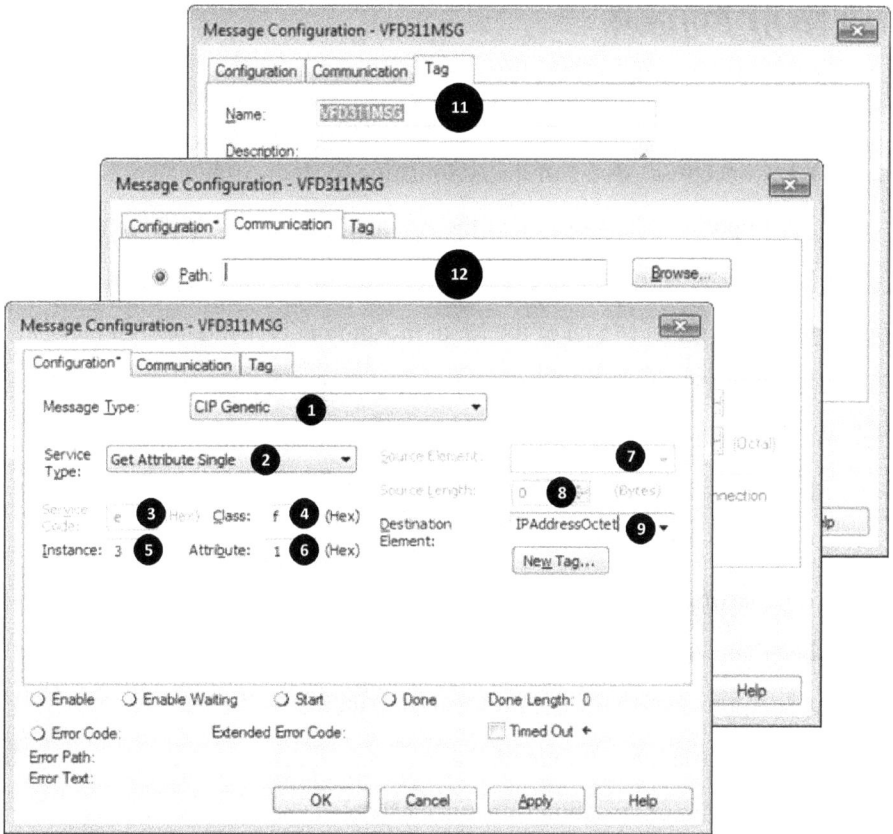

**Figure M.9: Ethernet IP Explicit Message Configuration**

## M.4 Exercise

i) Compare and contrast lab 6A and Lab 6B.

# List of Figures

[304]

[307]

# List of Tables